Internet of Things

Technology, Communications and Computing

The series Internet of Things - Technologies, Communications and Computing publishes new developments and advances in the various areas of the different facets of the Internet of Things. The intent is to cover technology (smart devices, wireless sensors, systems), communications (networks and protocols) and computing (theory, middleware and applications) of the Internet of Things, as embedded in the fields of engineering, computer science, life sciences, as well as the methodologies behind them. The series contains monographs, lecture notes and edited volumes in the Internet of Things research and development area, spanning the areas of wireless sensor networks, autonomic networking, network protocol, agent-based computing, artificial intelligence, self organizing systems, multi-sensor data fusion, smart objects, and hybrid intelligent systems.

Indexing: *Internet of Things* is covered by Scopus and Ei-Compendex **

Claudio Savaglio • Giancarlo Fortino •
MengChu Zhou • Jianhua Ma
Editors

Device-Edge-Cloud Continuum

Paradigms, Architectures and Applications

Springer

Editors
Claudio Savaglio
DIMES
Università della Calabria
Rende, Cosenza, Italy

Giancarlo Fortino
DIMES
Universita della Calabria
Rende, Cosenza, Italy

MengChu Zhou
New Jersey Institute of Technology
Newark, NJ, USA

Jianhua Ma
Hosei University
Tokyo, Japan

ISSN 2199-1073 ISSN 2199-1081 (electronic)
Internet of Things
ISBN 978-3-031-42193-8 ISBN 978-3-031-42194-5 (eBook)
https://doi.org/10.1007/978-3-031-42194-5

This Springer imprint is published by the registered company Springer Nature Switzerland AG
The registered company address is: Gewerbestrasse 11, 6330 Cham, Switzerland

Paper in this product is recyclable.

Contents

Toward the Edge-Cloud Continuum Through the Serverless Workflows

Christian Sicari, Alessio Catalfamo, Lorenzo Carnevale, Antonino Galletta, Antonio Celesti, Maria Fazio, and Massimo Villari

1 Introduction

In the last years, we have witnessed the rise of edge computing, a new trend that is in total opposition to cloud computing, aimed to collect and compute data as closely as possible to the data source. Even if edge computing has rapidly gained popularity, cloud has kept the leadership both for heavyweight jobs and data persistence because of the hardness on migration and integration. The gap between edge and cloud has been recently filled with an intermediate layer named fog, which is in charge of redirecting information to cloud and edge. The composition and orchestration of services between the three tiers have given rise to the cloud-edge Continuum (or just *Continuum*) [26] paradigm. Continuum's main goal consists of taking advantage of cloud, edge, and eventually even fog, to run applications where they best fit and readopting this, if something change in the environment or in the QoS parameters the applications is trying to satisfy [24, 30, 40].

Deploying software at the continuum is considered challenging for many reasons, such as architecture dependency, host federation, and global resource balancing, [28, 36]. However, the serverless paradigm has been recently introduced with the intent of making these problems surmountable. Serverless (i.e., function-as-a-service model) is a platform-independent approach for deploying and exposing services and relative APIs to final users, without worrying about the underlined system. Function-as-a-service (FaaS) engines are typically based on an orchestrator (i.e., Kubernetes) which is able to manage applications composed of many containers, load balance, and federate resources [24]. Serverless and FaaS paradigms are

C. Sicari (✉) · A. Catalfamo · L. Carnevale · A. Galletta · A. Celesti · M. Fazio · M. Villari
Department of Mathematics and Computer Sciences, Physical Sciences and Earth Sciences, University of Messina, Messina, Italy
e-mail: csicari@unime.it; alecatalfamo@unime.it; lcarnevale@unime.it; angalletta@unime.it; acelesti@unime.it; mfazio@unime.it; mvillari@unime.it

C. Savaglio et al. (eds.), *Device-Edge-Cloud Continuum*, Internet of Things,
https://doi.org/10.1007/978-3-031-42194-5_1

widely used in cloud-only applications, but thanks to their flexibility, some recent works are emerging with the purpose of deploying functions into the edge of the network for lightweight problems [3, 7, 33, 45]. For example, FaaS can be used for isolated and low-decoupled tasks, but it is not ideal for complex and coupled applications due to the impossibility of easily composing and integrating functions [26]. These drawbacks generate issues for continuum environments where, typically, applications are coupled in data-driven workflows with many tasks connected among different computing tiers [2, 20, 36].

In this chapter, we propose (i) new research guidelines for serverless orchestration in the continuum paradigm and (ii) a reference blueprint for the standard creation of a FaaS-based workflow orchestration. Specifically, we determine principles, definitions, a reference architectural model, and data structures that are useful for defining and orchestrating serverless workflows. Once the baseline is defined, we present a project called OpenWolf [42] as a ready-to-use solution for designing, deploying, and using serverless workflows, composed of many functions spread among the continuum. In order to evaluate the platform, we analyzed a deep learning application for image classification in a smart city scenario, considering five steps: collection, transformation, training, inference, and plotting.

The rest of the chapter is organized as follows. Section 2 describes the state of the art behind serverless and workflow, highlighting weaknesses and strengths of existing solutions. In Sect. 3 we describe the building blocks and the glossary term for any serverless-based workflow engine. In Sect. 4 we design a cloud-edge architecture used to manage and run serverless workflows. In Sect. 5 we describe OpenWolf, an open-source project compliant with the reference architecture. In Sect. 6 we describe a machine learning typical workflow using the glossary and the building block of this work; moreover, this workflow is tested using OpenWolf and the performances described below. Finally, in Sect. 7 we summarize the work presented and highlight the next research directions.

2 Background

The continuum aims to make a collaboration between the cloud and edge tiers in order to distribute near real-time processing on edge and massive processing on cloud [4]. Continuum faces several challenges related to different topics (i.e., security [41], scheduling) such that actual solutions [12] need to be reengineered to become suitable for the computing continuum.

Recently, serverless computing has emerged as a solution for distributing small functions using containers with the intent of reacting to external triggers (i.e., cronjobs, HTTP calls, message queue systems) [16, 46]. This new paradigm was well received by the scientific community, which tries to exploit it for orchestrating functions over the continuum [29, 35] by using different orchestrators, such as Kubernetes [6], Nomad [8, 27], and more [13]. Moreover, FaaS is used in the continuum to make the development, deployment, and automatic balancing easier, thanks to the underlined orchestrators [5, 31, 43].

The combined use of continuum and serverless pointed out the problem of composing functions, which means the capacity of concatenating functions for creating more complex applications. The authors [1] proposed three principles of serverless as (i) black-box functions, (ii) substitution, and (iii) double billing, which attempt to explain that composing FaaS application could be considered an anti-pattern. However, we do not agree with that statement.

The term workflow was used as a generic term for describing a well-defined organization of tasks connected in order to transform one or more inputs to a given output. In scientific literature, this term muted to *scientific workflows*, which is described [22] as a way to deal with data and pipelined computation steps in different application fields (i.e., bioinformatics, cheminformatics, ecoinformatics, geoinformatics, physics), without mastering a computer science background. For example, Kepler is a workflow grid-based, later extended [34] to support distributed computing on grid computing. Almost in parallel, the Pegasus system [9] was proposed to abstract the workflow as an ensemble of independent tasks. Such technology continued to have a progressive evolution, keeping the track of newer ones, such as grid [9, 21], Cloud [10], containers [19], and [44]. Going back to the last five years, workflows gained new popularity because of the increasing use of cloud computing and serverless. Indeed, the latter was widely adopted for designing and implementing workflows [17]. Perez et al. [32] designed a framework for executing Linux-based containers in a FaaS platform (i.e., AWS Lambda). Jiang et al. [17] integrated the scientific workflow into the main FaaS providers in order to exploit the serverless paradigm and make the implementation for end users (i.e., scientists) easier. Skyport [14] was instead a brilliant idea for creating black-box-based workflows, by means of an engine able to compose workflows as soft virtualized software (i.e., Docker containers). Recently, the workflow has become more sophisticated and accurate. It is not either a programming pattern or a software architecture design, but a computational on-premise engine where defining, storing, and deploying a composition of black-box functions [25]. Hyperstream [11] is a domain-specific tool used to deploy machine learning (ML) algorithms that are automatically fired by some incoming streaming data. One step ahead in this direction was moved in [18], where the authors proposed a workflow engine server (WES), which is a back-end engine used to store functions and workflows and run them when triggered by an event. Such an engine introduces workflow modularity and a validation schema, but it lacks integration with external systems and expandability with other functions. One of the most autonomous engines has been instead presented by Lopez et al. [23] with Triggerflow, a trigger-based orchestration of serverless workflows. It lacks a user-friendly workflow editor, a data schema for the functions, and an event global registry. However, Triggerflow has clear strengths, such as a mechanism to fire triggering-based workflow, an asynchronous communication channel, and a serverless model. A different approach for workflows was, instead, presented in [37], where the authors propose R-Pulsar, a cloud-edge engine that is able to trigger functions according to an interesting matching algorithm based on a decoupled associative message (AR) selection already presented in [38]. This helps in matching producers and consumers, as well as taking actions, such as running a function and starting a data production [39]. The abovementioned approaches prove good flexibility mostly when related to ML [15], but the utilization of serverless is still not totally well exploited.

3 Workflow Engine Characteristics and Principles

In this section, we put the stakes of the proposed workflow engine architecture, and we define the dictionary of terms that are used in the remainder of this paper, i.e., (i) state, (ii) event, (iii) workflow, and (iv) manifest

3.1 State

The main component of the architecture is the *state*. It mainly encapsulates a function and all the information related to it within a job. It is stateless, which means that the running job is not aware of other jobs interacting with it, and therefore the job behavior cannot change based on previous executions. As shown in Fig. 1, the job is composed of (i) *metadata* and (ii) a *function*. The latter is the code that includes the job's business logic, and it is encapsulated inside a container.

The metadata includes four different pieces of information, such as state description, handler instructions, input schema, and output schema. Specifically, they are described as follows:

Job description contains the job identifier, name, service description, and service class. They are used to quickly classify the service.

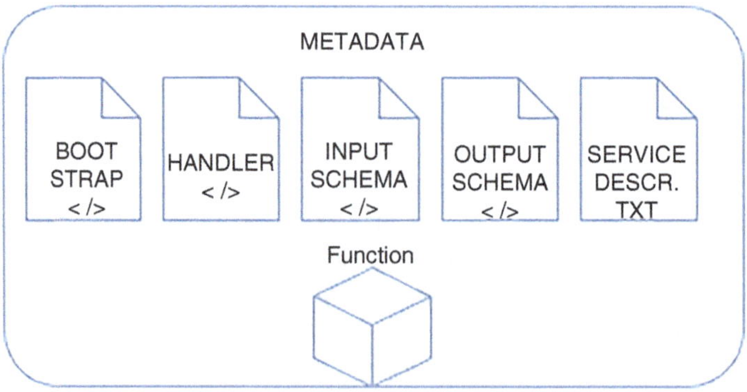

Fig. 1 The state encapsulates a function and all the information related to it within a job

Bootstrap instructions are run for instantiating a job inside the workflow engine. These could contain the code to either build an image, set the environment variables, or run a docker container.

Handler instructions are run every time a job is triggered. Basically, these validate the input schema, run a function using the passed and parsed parameters, wait for the function result, and finally parse results with a format compliant with the output schema.

Input/output schemas contain the schema of the acceptable input and the schema of the provided output. They are essential for creating compatible job chains.

Often, workflows also contain the *connectors*, a special kind of job that simply maps a job's output to the next jobs' input, according to their input/output schema. It is created on-premise during the workflow design, and it does not require an input and output predefined schema, since they change according to the workflow where they are located.

3.2 *Event*

An *event* is the only entity that can be processed in a workflow; it is originally sent from outside the workflow and then processed inside the workflow. All changes applied to an event are separately stored in a data lake, while the last version of the event is propagated through the workflows' jobs. An event is composed of both immutable and mutable data. The immutable data includes the following:

Event ID identifies the event uniquely, and it is managed directly by the workflow engine.

Workflow ID is a reference to the workflow which is processing/has processed the event.

The mutable data are generally updated by the workflow engine and by the jobs that process the event. This includes the following:

Status is a value in the following domain: ⟨Started, Processing, Error, Processed⟩.

Data is the last job's output.

Timestamp represents the date and time in which the last transformation has been completed.

3.3 *Workflow*

The workflow diagram shown in Fig. 2 represents how states interact between each other. The workflow starts when the first node is triggered by an external event, i.e.,

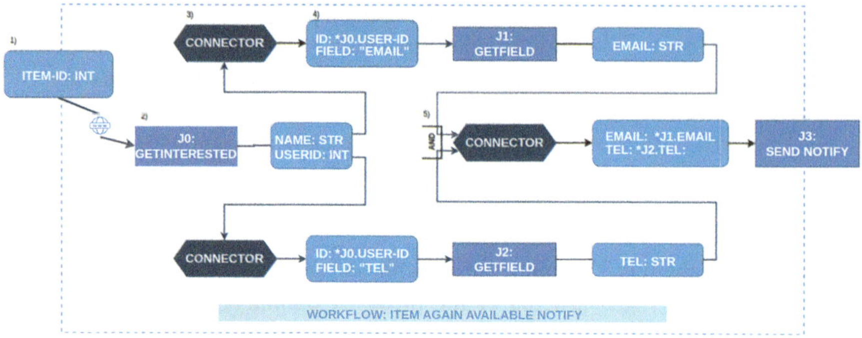

Fig. 2 Workflow example

the action 1 in Fig. 2, carrying on a data payload. Any event is directly connected to a *state* (action 2) and therefore to a *connector* (action 3). Connectors act as conciliators for filtering events with a specific state.

The first event is unique, and it is mapped one-to-one to a single workflow execution. This avoids overlaps with other events that follow the same workflow. Naturally, when an event passes through the *states*, it modifies its data according to the output of the previous state.

Within the workflow, any kind of links are allowed, such as many-to-many, many-to-one, and one-to-many. However, they must start and finish with only one *job*. When a many-to-one relationship (action 5 in Fig. 2) is defined, the triggering condition needs to be explained. In this regard, the condition may follow the Boolean algebra, i.e., using AND for combining two or more events that must be received before firing the next one, or using OR for combining two or more events according to the fact the only one of them is enough for firing the next state.

The workflow diagram shown in Fig. 2 is an example of e-commerce scenarios, where customers are notified both by an email and a short message system as soon as a product on which they are interested in is again available. Furthermore, the workflow is triggered by a web notification which says a given product is available again. The workflow fetches the users interested in this item using the *state* J0, J1, and J2 and then fetches the users' email and telephone number. Finally, the *state* J3 is used to notify the users. In this scenario, three connectors are used. Two connectors make compatible the J0's output with the J1 and J2's input; the last one maps instead the J1 and J2's output with the J3's input.

3.4 *Workflow Manifest*

In order to describe a workflow within a schema, we propose a manifest based on YAML format. The manifest, therefore, translates in processes what was designed, i.e., in Fig. 2.

Listing 1 Workflow Manifest Example

```
name: <workflow-name>
callbackUrl: <uri-where-to-send-result>
states:
  <state-id>:
    function:
      ref: <ref-to-function-id>
      config:
        key: value
    start: true
handlers:
  <handler-id>:
    endpoint: <endpoint-to-function>
    config:
      key: value
workflow:
  <state-id>:
    activation: <Boolean Equation>
    inputFilter: <jq command>
    outputFilter: <jq command>
```

As shown in the Listing 1, the manifest has (i) a name, (ii) a callback URL where the result is sent, and three more sections, such as (iii) states, (iv) handlers, and (v) workflow.

States list and describe all the status of the workflow. For each state, we define a name, handler, and global key-value configuration for the handler.

Handlers describe all the handlers called within the states. This attribute determines how to call the handler and the basic configurations that may be overwritten in the states' parts. The separation of states and functions sections allows using multiple times the same handler in different states.

Workflow describes how the states interact. For each state, we determine which previous states have triggered it and how to transform inputs and outputs. This part acts as a connector.

4 Architecture

The reference architecture for managing a serverless workflow is shown in Fig. 3. It is a four-layered architecture composed of (i) infrastructure, (ii) federation, (iii) serverless, and (iv) service layers. All layers are described as follows.

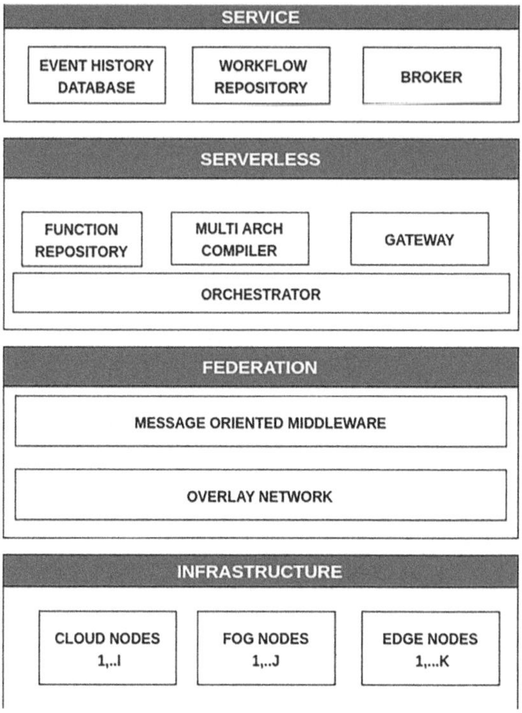

Fig. 3 Workflow engine architecture

The *infrastructure layer* contains the bare-metal nodes that are part of the continuum environment. Nodes may have different geographical locations, architecture characteristics, and distributions. The *federation layer* creates a communication interoperability among the nodes of the infrastructure layer. It is composed of an overlay network used to connect nodes with a message-oriented middleware (MOM), with the intent of exchanging data over the overlay itself. The *serverless layer* provides FaaS features to the underlined layer, i.e., the service layer. It uses a container orchestrator for deploying functions among the federation. It includes a function repository for storing the functions in the system, a compiler to build the same function in all the architecture available and compatible, and a gateway used to trigger the functions. The *service layer* is, instead, the top layer of the architecture. It adds capability of composition to the serverless layer. The service layer is composed of an event history database (EHD), a workflow repository, and a single agent. The EHD stores a permanent history of event transformation within the engine. Indeed, an event changes its mutable content when it is the input of a job. However, if a workflow is composed of n-jobs, the initial event will have n changes. Thus, the EHD stores all the n changes, along with the initial content. Furthermore, we had to consider a status history array field in the event data structure, as shown in Fig. 4. This approach allows to (i) keep track of the event history, (ii) keep track of the event transformation, (iii) log every change, and (iv) recover any workflow state. The *workflow repository* stores the manifest files that contain the workflow

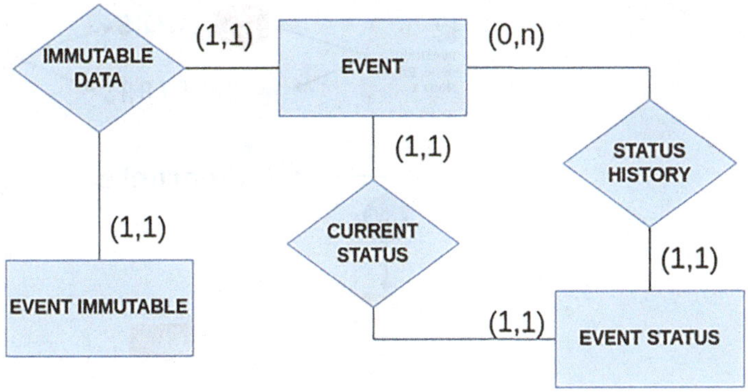

Fig. 4 Event data model

descriptions according to the structure defined in Sect. 3.4. The *broker* coordinates the service layer and, more in general, the overall infrastructure. It is basically in charge of receiving the external events and intercepting the execution of a function inside a triggered state, recognizing that it uses the proper workflow manifest in the workflow repository, and then updates the EHD for saving the actual data coming from the events or from the states.

5 OpenWolf: Serverless Workflow Engine

The architecture shown in Fig. 3 is implemented in an open-source project currently under development, called OpenWolf [42]. The OpenWolf architecture is shown in Fig. 5, and it is composed of four main elements: (i) Kubernetes, (ii) OpenFaaS, (iii) Redis, and (iv) the OpenWolf agent.

Kubernetes works between the federation and serverless layers. It is used to federate continuum nodes using an own defined overlay network. Moreover, it also provides the orchestration tools needed by the serverless layer for deploying functions among the continuum. OpenFaaS works at the serverless layer as the engine to store, compile, deploy, and manage functions in conjunction with Kubernetes. Redis works inside the service layer, and it acts as EHD and workflow repository. Indeed, it stores the workflow manifests, but it also keeps track of the workflow executions. For the latter, OpenWolf uses a well-defined event structure expressed using a JSON format, in which the main properties are called *ctx* and *data*. The first one represents the event context, and it is composed of the *workflowID*, which references the workflow to which event it belongs; the *execID*, which distinguishes the different executions of the same workflow; and the *state*, which references the state that has returned the event. The *data* property, instead, is the function's output

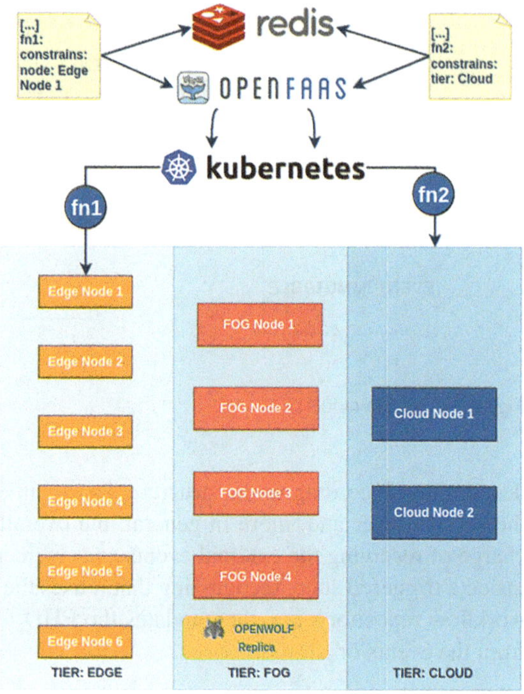

Fig. 5 Workflow engine architecture

itself, and, unlike the *ctx* that is read and set by the workflow agent, this is fully managed by the function.

An event example is proposed in the Listing 2, which is fired by *State* C in the workflow shown in Fig. 6.

Listing 2 Event Data Structure

```
{
    "ctx": {
        "workflowID": "inference-traffic",
        "execID": inference-traffic.123,
        "state": "C"
    },
    "data": {
        "AIQ": 47,
        "Scale": "EU"
    }
}
```

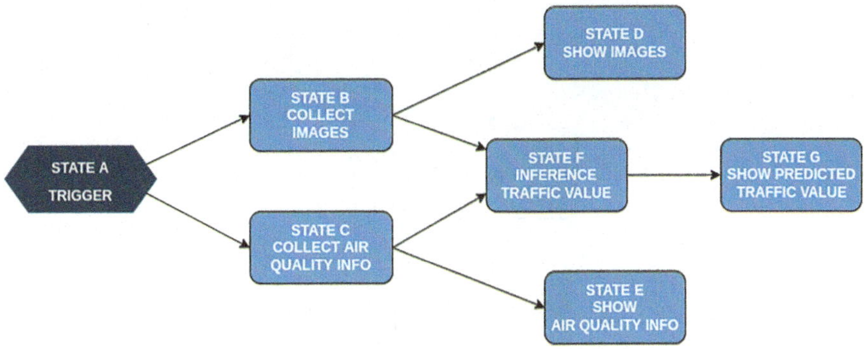

Fig. 6 Example of workflow in data analytics

OpenWolf agent acts as a broker for the workflow statuses, as it is used to achieve the function composition feature. OpenWolf ensures that any event will follow the correct path in the workflow and triggers the correct states in the workflow with a proper transformation of the right income event. In this regard, the OpenWolf agent is deployed as a standalone stateless microservice inside the same Kubernetes cluster used to run the serverless functions. The agent exposes two interfaces. The first one is a public interface used to trigger a workflow from the external. The second one is closed inside the Kubernetes cluster, and it is used as a callback URL for each asynchronous function triggered by any workflow. By doing that, the agent intercepts all the events belonging to a workflow, extracts the context information, and uses it for fetching all the workflow and current execution information. Therefore, it triggers the next states in the manifest, forwarding the right received event with the updated *ctx* property. This process is described more concisely in the activity diagram in Fig. 7.

6 Use Case

Smart cities are a typical scenario for a computing continuum use case. For example, in private and public spaces, we could find Internet of Things (IoT) sensors and small computing devices, like cameras and Raspberry Pi for monitoring buildings, traffic, or environment parameters. These data are then typically processed in local data factories provided by private citizens, municipalities, or research institutes, and often they trust on private cloud providers like AWS or Azure, i.e., for long-time storage or processing. As a consequence, it is easy to find the three continuum layers over them. In the following, we will analyze a typical pipe for image processing. Smart cities rely on this kind of algorithm for detecting violent and dangerous situations, traffic rule violation, or roadside surveillance application (Fig. 8).

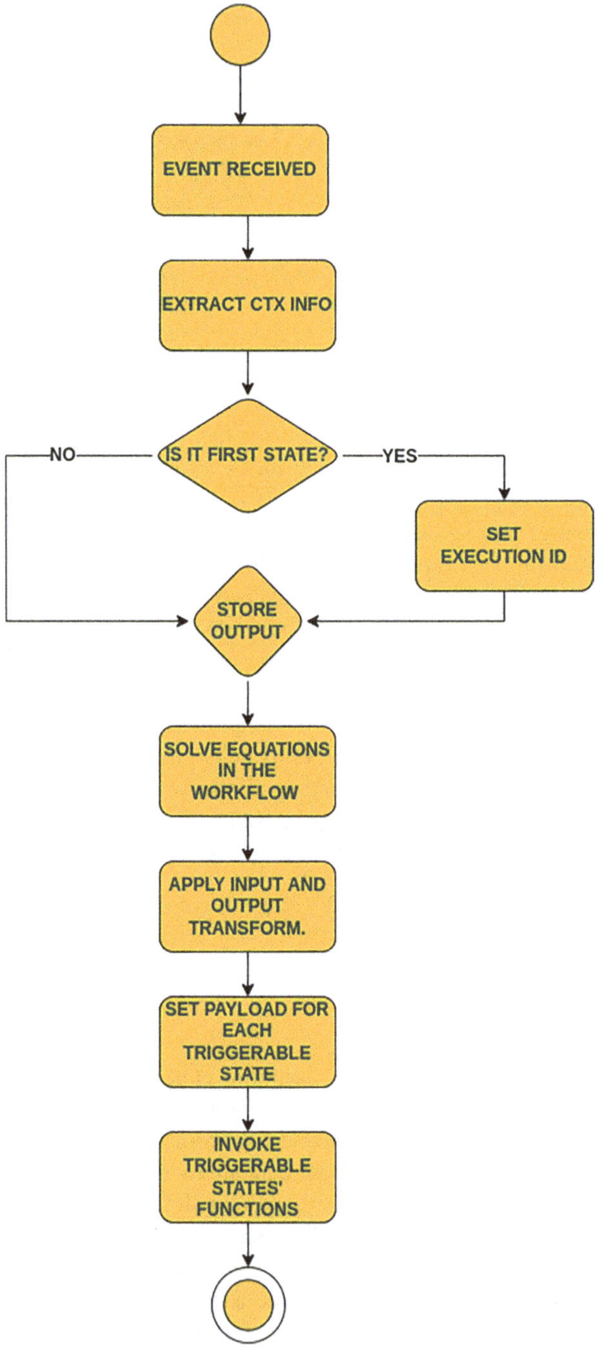

Fig. 7 OpenWolf agent actions

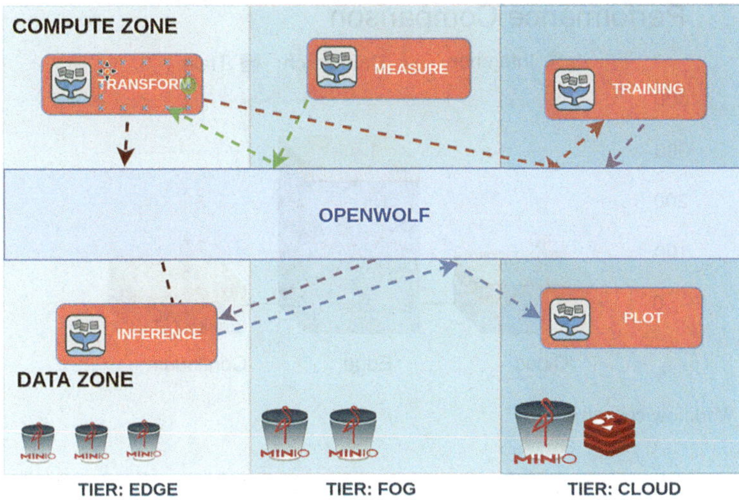

Fig. 8 OpenWolf for image processing

The designed image processing workflow is composed of five states. Each state represents a function, i.e., a process inside the workflow. Each state is deployed within one of the computing continuum tiers according to the static scheduling rule defined in the workflow manifest. According to the states' descriptions:

Collect exploits a camera stream for collecting environment images.

Transform edits the images, cleaning and filtering noisy data. It can be run on any of the continuum's tiers.

Train trains a recurrent neural network (RNN) model used to analyze the collected images.

Inference predicts the input image's label using the latest model produced by the train state.

Show pushes the result of the inference over a web page.

The first problem we identify on continuum, mostly when FaaS is implemented, is having a good scheduler for deploying functions according to specific quality of services (QoS), i.e., latency, network bandwidth usage, and resource performances. The second problem, which is relied on the first one, is where to put the data. These typically are collected on the edge, but they could be only partially computed on the edge or delivered to the cloud for massive analysis. QoS are directly dependent on the service we are providing in the smart city. For example, road traffic monitoring could require optimizing accuracy, whereas shotgun detection could require real-time analysis. Our proposed solution aims to give the possibility to directly customize what and where data are processed, trying to satisfy any kind of QoS.

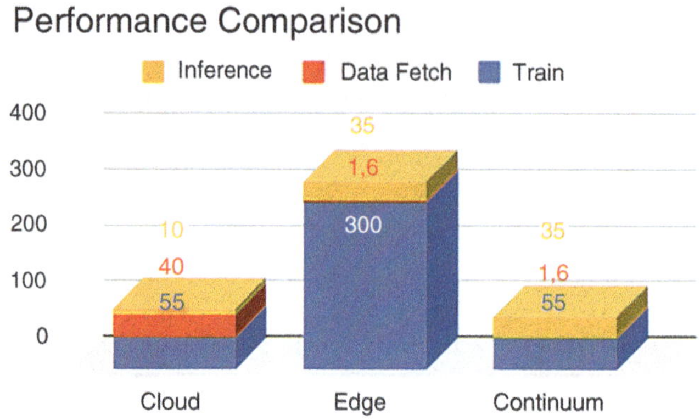

Fig. 9 Workflow comparison

6.1 *Performance Evaluation*

We evaluated the workflow in three different environments. We considered three key workflows' moments: (i) training, (ii) data fetching, and (iii) data inference in a full-cloud, a full-edge, and a continuum test bed. In the latter case, cloud nodes were in charge of the model training, whereas edge nodes were focused on collecting and inferencing data. These three functions have been encapsulated inside three different OpenFaaS functions. The dataset adopted is the CIFAR-10, the algorithm was trained for 50 epochs, and the implementation of the RNN was achieved with PyTorch. The data training size is 130 MB, while the data test size, used during the inference, is around 100 MB. As an edge node, we used a single Raspberry Pi 4, with an ARM64 operating system, a 4 GB of RAM, and a 1.5 GHz quad-core processor. As a cloud node, we used a virtual machine with 16 GB of RAM, a 2.8 GHz quad-core processor, and a x64 operating system.

Results are shown in Fig. 9. The edge node needed 400% of the cloud performance for training, and 120% for inferencing data, but the time to use local data is close to zero. On the other hand, cloud requires 45 seconds for transferring data from the edge object storage to the local storage. Moreover, edge device does not require network usage, but cloud will use WAN network for receiving the entire test dataset from the edge object storage. Finally, distributing the computation over the continuum allows exploiting the cloud training time and the edge data locality, avoiding any massive network usage. Unfortunately, as a direct consequence, the inference is done inside the edge, but as shown in Fig. 9, the overall performance in the continuum is better than both the edge and the cloud.

7 Conclusion and Future Works

In the era of serverless and microservice architecture, workflows are slowly going to gain popularity as a tool to mix serverless services and to deploy them in order to compose complex functions in modern engine infrastructure based on the cloud.

Historically, workflows are recognized as a computation chain where the processes involved depend on the specific field where they are acted. In the last two years, this term instead has started to appear in different fields, like microservices, FaaS, and cloud-edge continuum. In some way, the scientific community shares the idea that workflows allow enabling the cooperation between functions, services, and in general network hosts.

This trend is fully reasonable since we managed to deploy functions and services everywhere, just to think to the new concept of "Internet of Everything." However, we did not manage to link these capabilities together. To try to reach this scope, different open-source and enterprise providers proposed different "linking services," which have been called FaaS orchestrators. These are of course valid products, but do not trust on a standard and are not integrable, and each of them does not absolve at all to all the requirements a function workflows could ask.

In this scenario, we started from scratch, defining the workflow concept. Therefore, we firstly determined what are the elements involved in workflows, i.e., jobs, events, and how they are related together. After that, we defined a designing schema for workflows with clear terms, figures, and data models. Finally, using these tools, we proposed a reference architecture for the management of a workflow platform over a continuum cluster. After the introduction of those glossary terms and architecture patterns, we validate our work presenting OpenWolf, a recently born open-source engine, used to design and develop FaaS workflows for heterogeneous Kubernetes clusters, and we measured its capabilities considering a continuous learning workflow applied to a smart city environment, used to keep under security control a square or a street.

This work can be considered a starting point for the serverless workflow field, but we still have to deal with different challenges, like (i) designing the security aspects on the engine, (ii) designing the fault tolerance aspects on the nodes, and (iii) implementing a workflow engine that is able to respect this reference architecture. All these challenges will be faced up in the next future works, with the main intent of providing a usable prototype of a workflow engine platform.

References

1. I. Baldini, P. Cheng, S. Fink, N. Mitchell, V. Muthusamy, R. Rabbah, P. Suter, O. Tardieu, The serverless trilemma: function composition for serverless computing, in *Proceedings of the 2017 ACM SIGPLAN International Symposium on New Ideas, New Paradigms, and Reflections on Programming and Software* (2017), pp. 89–103. https://doi.org/10.1145/3133850.3133855

2. D. Balouek-Thomert, E. Renart, A. Zamani, A. Simonet, M. Parashar, Towards a computing continuum: enabling edge-to-cloud integration for data-driven workflows. Int. J. High Perform. Comput. Appl. **33**, 1159–1174 (2019).
3. L. Baresi, D. Filgueira Mendonça, Towards a serverless platform for edge computing, in *2019 IEEE International Conference on Fog Computing (ICFC)* (2019), pp. 1–10
4. L. Bittencourt, R. Immich, R. Sakellariou, N. Fonseca, E. Madeira, M. Curado, L. Villas, L. Silva, C. Lee, O. Rana, The internet of things, fog and cloud continuum: integration and challenges. Internet Things **3–4**, 135–155 (2018). https://arxiv.org/abs/1809.09972
5. A. Bocci, S. Forti, G. Ferrari, A. Brogi, Type, pad, and place: avoiding data leaks in Cloud-IoT FaaS orchestrations, in *2022 22nd IEEE International Symposium on Cluster, Cloud and Internet Computing (CCGrid)* (2022), pp. 798–805
6. R. Boutaba, *The Cloud to Things Continuum*, Association for Computing Machinery, Virtual Event, Canada (2021). https://dl.acm.org/doi/10.1145/3501255.3501407
7. M. Ciavotta, D. Motterlini, M. Savi, A. Tundo, DFaaS: decentralized function-as-a-service for federated edge computing, in *2021 IEEE 10th International Conference on Cloud Networking (CloudNet)* (2021), pp. 1–4
8. I. Cilic, I. Zarko, M. Kusek, Towards service orchestration for the cloud-to-thing continuum, in *2021 6th International Conference on Smart and Sustainable Technologies, SpliTech 2021* (2021)
9. E. Deelman, J. Blythe, Y. Gil, C. Kesselman, G. Mehta, S. Patil, M. Su, K. Vahi, M. Livny, Pegasus: mapping scientific workflows onto the grid. Lecture Notes Computer Science (Including Subseries Lecture Notes Artificial Intelligence Lecture Notes Bioinformatics) **3165**, 11–20 (2004)
10. E. Deelman, K. Vahi, M. Rynge, G. Juve, R. Mayani, R. Silva, Pegasus in the cloud: science automation through workflow technologies. IEEE Internet Comput. **20**, 70–76 (2016)
11. T. Diethe, M. Kull, N. Twomey, K. Sokol, H. Song, M. Perello-Nieto, E. Tonkin, P. Flach, HyperStream: a workflow engine for streaming data (2019). http://arxiv.org/abs/1908.02858
12. S. Dustdar, V. Pujol, P. Donta, On distributed computing continuum systems. IEEE Trans. Knowl. Data Eng. **XX**, 1–14 (2022)
13. N. Faria, D. Costa, J. Pereira, R. Vilaça, L. Ferreira, F. Coelho, AIDA-DB: a data management architecture for the edge and cloud continuum, in *2022 IEEE 19th Annual Consumer Communications & Networking Conference (CCNC)* (2022), pp. 1–6
14. W. Gerlach, W. Tang, A. Wilke, D. Olson, F. Meyer, Container orchestration for scientific workflows, in *Proceedings – 2015 IEEE International Conference on Cloud Engineering, IC2E 2015* (2015), pp. 377–378
15. Z. Houmani, D. Balouek-Thomert, E. Caron, M. Parashar, Enabling microservices management for deep learning applications across the Edge-Cloud Continuum, in *2021 IEEE 33rd International Symposium on Computer Architecture and High Performance Computing (SBAC-PAD)* (2021), pp. 137–146
16. A. Jangda, D. Pinckney, Y. Brun, A. Guha, Formal foundations of serverless computing. Proc. ACM Program. Lang. **3**, 1–26 (2019). https://dl.acm.org/doi/10.1145/3360575
17. Q. Jiang, Y. Lee, A. Zomaya, Serverless execution of scientific workflows. Lecture Notes Computer Science (Including Subseries Lecture Notes Artificial Intelligence Lecture Notes Bioinformatics), LNCS, vol. 10601, pp. 706–721 (2017)
18. A. Jasinski, Y. Qiao, J. Keeney, E. Fallon, R. Flynn, A workflow engine server for the design of adaptive and scalable workflows, in *30th Irish Signals and Systems Conference, ISSC 2019* (2019)
19. D. Kimovski, R. Mathá, J. Hammer, N. Mehran, H. Hellwagner, R. Prodan, Cloud, fog, or edge: where to compute?. IEEE Internet Comput. **25**, 30–36 (2021)
20. D. Kimovski, C. Bauer, N. Mehran, R. Prodan, Big data pipeline scheduling and adaptation on the computing continuum, in *2022 IEEE 46th Annual Computers, Software, and Applications Conference (COMPSAC)* (2022), pp. 1153–1158
21. K. Lee, N. Paton, R. Sakellariou, E. Deelman, A. Fernandes, G. Mehta, Adaptive workflow processing and execution in Pegasus, in *2008 The 3rd International Conference on Grid and Pervasive Computing – Workshops* (2008), pp. 99–106

22. B. Ludäscher, I. Altintas, C. Berkley, D. Higgins, E. Jaeger, M. Jones, E. Lee, J. Tao, Y. Zhao, Scientific workflow management and the KEPLER system. Concurrency Comput. Prac. Exp. **18**, 1039–1065 (2006)
23. P. López, A. Arjona, J. Sampé, A. Slominski, L. Villard, Triggerflow: trigger-based orchestration of serverless workflows, in *DEBS 2020 – Proceedings of the 14th ACM International Conference on Distributed and Event-Based Systems* (2020), pp. 3–14
24. A. Luckow, K. Rattan, S. Jha, Pilot-edge: distributed resource management along the edge-to-cloud continuum, in *2021 IEEE International Parallel and Distributed Processing Symposium Workshops (IPDPSW)* (2021), pp. 874–878
25. M. Malawski, A. Gajek, A. Zima, B. Balis, K. Figiela, Serverless execution of scientific workflows: experiments with HyperFlow, AWS Lambda and Google Cloud Functions. Futur. Gener. Comput. Syst. **110**, 502–514 (2020). https://www.sciencedirect.com/science/article/pii/S0167739X1730047X
26. X. Masip-bruin, E. Marín-tordera, S. Sánchez-lópez, J. Garcia, A. Jukan, A. Ferrer, A. Queralt, A. Salis, A. Bartoli, M. Cankar, C. Cordeiro, J. Jensen, J. Kennedy, Managing the cloud continuum: lessons learnt from a real fog-to-cloud deployment. Sensors **21**(9) (2021). https://www.mdpi.com/1424-8220/21/9/2974
27. X. Merino, C. Otero, D. Nieves-Acaron, B. Luchterhand, Towards orchestration in the cloud-fog continuum, in *Conference Proceedings – IEEE SOUTHEASTCON*, vol. 2021 (2021)
28. H. Mueller, S. Gogouvitis, H. Haitof, A. Seitz, B. Bruegge, Poster abstract: continuous computing from cloud to edge, in *2016 IEEE/ACM Symposium on Edge Computing (SEC)* (2016), pp. 97–98 (2016)
29. D. Mukherjee, D. Pal, P. Misra, Workflow for the Internet of Things, in *ICEIS 2017 - Proceedings of the 19th International Conference on Enterprise Information Systems*, vol. 2, Porto, Portugal, April 26–29 (2017)
30. A. Morichetta, V. Pujol, S. Dustdar, A roadmap on learning and reasoning for distributed computing continuum ecosystems, in *2021 IEEE International Conference on Edge Computing (EDGE)* (2021), pp. 25–31
31. E. Paraskevoulakou, D. Kyriazis, Leveraging the serverless paradigm for realizing machine learning pipelines across the edge-cloud continuum, in *2021 24th Conference on Innovation in Clouds, Internet and Networks and Workshops (ICIN)* (2021), pp. 110–117
32. A. Pérez, G. Moltó, M. Caballer, A. Calatrava, Serverless computing for container-based architectures. Futur. Gener. Comput. Syst. **83**, 50–59 (2018)
33. T. Pfandzelter, D. Bermbach, tinyFaaS: a lightweight FaaS platform for edge environments, in *2020 IEEE International Conference on Fog Computing (ICFC)* (2020), pp. 17–24
34. M. Płóciennik, T. Zok, A. Gómez-Iglesias, F. Castejón, A. Bustos, M. Rodríguez-Pascua, J. Velasco, Workflows orchestration in distributed computing infrastructures, in *Proceedings of the 2012 International Conference on High Performance Computing and Simulation, HPCS 2012* (2012), pp. 616–622
35. P. Raith, S. Nastic, S. Dustdar, Serverless edge computing – where we are and what lies ahead. IEEE Internet Comput. **27**(3), 50–64 (2023)
36. A. Ranjan, F. Guim, M. Chincholkar, P. Ramchandran, R. Mishra, S. Ranganath, Convergence of edge services & edge infrastructure. *2021 IEEE Conference on Network Function Virtualization and Software Defined Networks (NFV-SDN)* (2021), pp. 96–99
37. E. Renart, D. Balouek-Thomert, M. Parashar, An edge-based framework for enabling data-driven pipelines for IoT systems, in *2019 IEEE International Parallel and Distributed Processing Symposium Workshops (IPDPSW)* (2019), pp. 885–894
38. E. Renart, J. Diaz-Montes, M. Parashar, Data-driven stream processing at the edge, in *2017 IEEE 1st International Conference on Fog and Edge Computing (ICFEC)* (2017), pp. 31–40
39. E. Renart, D. Balouek-Thomert, X. Hu, J. Gong, M. Parashar, Online decision-making using edge resources for content-driven stream processing, in *2017 IEEE 13th International Conference on E-Science (e-Science)* (2017), pp. 384–392
40. D. Rosendo, A. Costan, G. Antoniu, M. Simonin, J. Lombardo, A. Joly, P. Valduriez, Reproducible performance optimization of complex applications on the edge-to-cloud continuum, in *2021 IEEE International Conference on Cluster Computing (CLUSTER)* (2021), pp. 23–34

41. A. Ruggeri, A. Celesti, M. Fazio, M. Villari, An innovative blockchain-based orchestrator for osmotic computing. J. Grid Comput. **20**, 1–17 (2022)
42. C. Sicari, L. Carnevale, A. Galletta, M. Villari, OpenWolf: a serverless workflow engine for native cloud-edge continuum, in *He 7th IEEE Cyber Science and Technology Congress (CyberSciTech 2022)* (2022)
43. C. Smith, A. Jindal, M. Chadha, M. Gerndt, S. Benedict, FaDO: FaaS functions and data orchestrator for multiple serverless edge-cloud clusters, in *2022 IEEE 6th International Conference on Fog and Edge Computing (ICFEC)* (2022), pp. 17–25
44. K. Vahi, M. Rynge, G. Papadimitriou, D. Brown, R. Mayani, R. Silva, E. Deelman, A. Mandal, E. Lyons, M. Zink, Custom execution environments with containers in Pegasus-enabled scientific workflows, in *2019 15th International Conference on EScience (eScience)* (2019), pp. 281–290
45. B. Vincenzo et al., Disclosing edge intelligence: a systematic meta-survey. Big Data Cogn. Comput. **7**(1), 44 (2023)
46. J. Wen, X. Jin, Rise of the planet of serverless computing: a systematic review; Rise of the planet of serverless computing: a systematic review. ACM Trans. Softw. Eng. Methodol. **32**, 1–61 (2023). https://dl.acm.org/doi/10.1145/3579643

Firmware Dynamic Analysis Through Rewriting

Claudia Greco, Michele Ianni, Antonella Guzzo, and Giancarlo Fortino

1 Introduction

The spread of Internet of Things (IoT) devices and their full integration into everyday life is one of the major factors defining the current technology landscape. With the embedding of computational power and persistent connectivity to the Internet, an ever-increasing number of everyday objects are now IoT devices and created new levels of automation, efficiency, and convenience. IoT devices are being used in a wide range of applications and ecosystems, including smart homes, healthcare, transportation, industrial settings, and daily living [1, 2]. By gathering and transmitting data, smart objects are enabling new possibilities for innovation and improvement. Considering their constant use and the access they have to our data, ensuring that these devices are safe is an urgent concern, exacerbated by the fact that they still lack of adequate security and safety measures, putting privacy at risk and making IoT devices increasingly appealing targets for attackers [3]. The vulnerabilities present in IoT devices make them highly susceptible to attacks, and they are frequently viewed as low-hanging fruit by malicious actors, owing to their ease of exploitation [4]. The necessity of conducting a thorough, security-focused evaluation of IoT devices has been well-established [5]. However, conventional analysis methods are often not suitable for the IoT environment: the dynamic analysis of the firmware of these devices typically requires that code is not executed in the device's native execution environment, but in a controlled one. The reason behind this is manifold. First of all, the use of dynamic analysis on the native device may require expensive hardware which may not be readily available to the analyst

C. Greco (✉) · M. Ianni · A. Guzzo · G. Fortino
Department of Computer Science, Modeling, Electronics and System Engineering (DIMES), University of Calabria, Arcavacata, Italy
e-mail: claudia.greco@dimes.unical.it; michele.ianni@unical.it; antonella.guzzo@unical.it; giancarlo.fortino@unical.it

C. Savaglio et al. (eds.), *Device-Edge-Cloud Continuum*, Internet of Things,
https://doi.org/10.1007/978-3-031-42194-5_2

and that may be prone to damage during testing. Also, it can be hard to supply inputs to guide analysis (as happens with fuzzing) and debugging. As explained in [6] using fuzzing, it is hard to detect memory corruption vulnerabilities on low-cost bare-metal devices lacking security mechanisms, because of the lack of visible effects. While, in theory, the possibility of using the device's ports for debugging exists, these ports are often obscured or inaccessible. Additionally, with physical hardware, it is not feasible to perform concurrent executions, which is essential for dynamic analysis.

For this purpose, we rely on the emulation of the firmware, better known as *firmware re-hosting*. This process separates the firmware from the hardware and emulates it to run on a different architecture without the need for actual hardware. Firmware re-hosting offers several benefits, including the ability to execute in a controlled environment, use debuggers for greater insight, concentrate solely on software components, and benefit from scalability. However, that of firmware re-hosting is a challenging task due to the fact that the firmware frequently retrieves input directly from the device peripherals, which can have their own unique access definitions and different configurations and interfaces.

Several solutions have been proposed in literature for firmware re-hosting, based on approaches such as hardware-in-the-loop (HITL), low-level abstractions, learning, or symbolic execution. Despite the significant progress that has been made in the area of automated re-hosting and analysis of firmware, the current solutions come with drawbacks. While the hardware-in-the-loop approach has its advantages, it is often not feasible due to the difficulty or impossibility of obtaining real hardware, or the risk of damaging expensive components during large-scale automated analysis. Approaches based on abstractions usually incorporate binary instrumentation techniques to intercept calls to functions that interact with the hardware. Binary instrumentation adds a substantial overhead to an already slow emulation environment and severely impacts the performance of dynamic vulnerability discovery techniques like fuzzing, which requires a high number of executions of binary code to take place.

This chapter has different goals: we analyze the current solutions adopted in literature to enable dynamic analysis on the re-hosted firmware and their limitations. We offer our point of view on how progress can be made in this context in order to enable faster vulnerability discovery processes by using traditionally employed techniques such as fuzzing. In particular, we discuss our idea of replacing the device peripherals and the interactions between the firmware and hardware with high-level operations, bringing all the firmware functioning at the software level. This chapter directly addresses a crucial aspect within the broader domain of the device edge cloud continuum. By exploring the security challenges and vulnerabilities of IoT devices, specifically focusing on firmware re-hosting and dynamic analysis, our research aligns with the overarching theme of advancing paradigms, architectures, and applications in this interconnected landscape.

The chapter is organized as follows: in Sect. 2, we provide background on the concept of firmware re-hosting and its necessity for dynamic analysis, along with basic notions about vulnerability discovery well-known techniques and different levels of analysis. In Sect. 3, we discuss hardware emulation, its challenges, and the limitations of the existing approaches. In Sect. 4, we review the state of the art in firmware emulation and discuss the drawbacks of existing solutions. Finally, in Sect. 5, we describe the idea behind our proposal.

2 Background

Program analysis plays a crucial role when dealing with the security assessment of software systems, and, over the years, there has been significant progress in the development of new techniques and methodologies to accomplish this task. Great effort has been put to make the process of program analysis scalable, leading to the development of dynamic analysis tools such as fuzzers and symbolic execution engines that allow to analyze binary programs that, as often occurs, do not come with their source code. Popular examples in the security community of tools for fuzzing and symbolic execution are AFL [7] and angr [8]. However, program analysis tools, especially when performing dynamic analysis, need high levels of parallelism and scalability to function properly, which necessitates moving the execution into an emulated environment.

With the advent of IoT, the use of such tools, firstly meant to be applied on desktop and mobile systems, has been extended to various devices and their firmware. The necessity to shift the execution to a virtual environment in the case of firmware is supported by its dependency on the hardware, which limits the ability to observe its behavior. For example, in order to perform security testing of a firmware by means of fuzzing, it would be often necessary to substitute the values derived from peripheral interactions with inputs generated by a fuzzer. It follows the execution must be taken out of its native environment and performed into an emulated and controlled environment, without involving physical components. The process of emulating a firmware in a way that accurately replicates its behavior on real hardware is referred to as *re-hosting*. Firmware re-hosting allows to thoroughly examine and manipulate firmware in manners that are not feasible on physical hardware and offers many benefits to analysts, including the ability to limit the scope of analysis to software components alone while providing scalability. In this way, no physical embedded component of the device is necessary for the security analysis process to be held, and the program execution can be attached to debuggers, making it possible to get a more in-depth understanding of program execution. Unfortunately, the task of firmware re-hosting is not without challenges, since while running the firmware interacts with peripherals. It follows that in order to gain a properly functioning emulation of the embedded system, firmware re-hosting implies modeling, along the CPU, the behavior of the device peripherals. Peripherals fall in two categories: *on-chip* peripherals include components such as timers, bus controllers, networking elements, and serial ports, while *off-chip* peripherals include sensors, actuators, external storage devices, and other circuit board circuitry that is accessed via on-chip peripherals. Specifically, on-chip peripherals, such as general-purpose input/output (GPIO) or bus interfaces like inter-integrated circuit (I2C) and serial peripheral interface (SPI), intermediate the firmware and off-chip peripheral communications and are typically controlled by the CPU through memory-mapped input/output (MMIO), allowing programs to access them via memory. The absence of these components can result in the firmware crashing or producing outcomes that deviate from those generated when real hardware is employed, and since a good many of system functions involve interactions with both on-chip and off-chip peripherals, realizing a proper emulation of them is vital.

2.1 Vulnerability Discovery Techniques

As we stated, there are some popular dynamic analysis techniques employed during the vulnerability discovery process of a system that benefit from its emulation. In this section, we briefly outline some of the most popular dynamic analysis methodologies.

2.1.1 Fuzzing

The goal of fuzzing is to identify inputs that cause the program to behave in unexpected ways or even crash, revealing the presence of vulnerabilities, bugs, or other security issues in the programs that could be exploited by attackers. During fuzzing, a large number of values randomly or semi-randomly generated by a *fuzzer* are given as input to the target program, which is then monitored for possible crashes, errors, or unexpected outputs. Popular fuzzers are AFL [7], libFuzzer,[1] and Honggfuzz.[2]

2.1.2 Concolic Execution

combines symbolic execution and concrete execution to analyze the behavior of computer programs. It runs the program with concrete inputs while maintaining a symbolic state. It is also known as *dynamic symbolic execution*, and it is different from static symbolic execution as it explores only one path at a time, determined by the concrete inputs. To explore different paths, the technique "flips" path constraints and uses a constraint solver to calculate concrete inputs that result in alternative branches.

[1] https://llvm.org/docs/LibFuzzer.html

[2] https://github.com/google/honggfuzz

2.1.3 Binary Instrumentation

is used to gather information about the behavior and performances of an executable by adding code to the compiled binary. Instrumentation is meant to track and record data about program execution, such as function call information, memory accesses, and performance metrics, which can be used for a variety of purposes, such as debugging, testing, profiling, and other security analysis. The instrumentation process is performed at the machine code level, making it platform-independent, and can be used to analyze a wide range of software, including low-level system code, firmware, and high-level applications. However, binary instrumentation can also introduce overhead at the program execution, as the added code can slow down the program and consume more memory.

2.2 Analysis Levels

The software security analysis of an embedded system can be performed at different levels: full-system, process level, or application level.

- With **full-system emulation**, the firmware is meant to run inside a virtual environment recreated by an emulator, which is supposed to mimic any component of the original system, from the processor to the hardware peripherals. This approach is able to test the system in every aspect, allowing to generate the same data and behavior of the original target system, but it is the hardest to achieve, as well as slower when compared to other levels of emulation. Full-system emulation is possible by means of base emulators such as QEMU and Simics, although they only provide a small set of peripherals, which does not cover the wide and diverse range of possible hardware in embedded systems.
- The analysis at **process level** allows the emulation of specific processes' behavior inside the target system. The execution of the processes can be performed inside the native system or a different hardware platform with an operating system providing an execution environment that resembles the native one. Emulators such as QEMU and Simics allow the process level analysis though the user mode emulation. Process level emulation is faster than the full-system emulation; however, the results of the analysis may differ from the reality if the emulated execution environment is not faithful to the original, thus compromising the vulnerability discovery process.
- Analysis at the **application level** consists in analyzing the single application that can run in the native target system. This can be done both statically, by extracting application-specific data, and dynamically, by running the application itself. The limitation of the static approach is that reducing the analysis to the evaluation of statically extracted data can detect existing vulnerabilities in the specific application, but not within the system that interacts with that application. In the case of dynamic analysis, the execution is usually carried in the native

execution environment. This analysis level is faster than the others, since it does not require the emulation of the system or process; however, the emulation may not be accurate if the target application depends on native hardware features not supported by the execution environment.

A problem encountered when full-system emulation is not involved is that the host platform used to run the firmware is not necessarily capable of supporting the use of dynamic analysis tools. This results from the fact that IoT systems are generally lightweight and have reduced computing and storage resources, compared to traditional systems.

3 Motivation

Firmware re-hosting in IoT devices constitutes a significant challenge due to the wide heterogeneity involved in both hardware and software components, especially in comparison to desktop and mobile systems where the standardized execution environments and limited number of operating systems and architectures make the issue much easier to handle. The creation of a single generic emulator that is capable of hosting transparently a certain firmware turns out to be a highly impractical goal due to the remarkable diversity of existing embedded systems and architecture designs, as well as the proprietary nature of some chip designs. This diversity results from the combination of various hardware architectures (x86, ARM, MIPS, and so on), different types of embedded peripherals, multiple operating systems, and customized configurations and interfaces. The conjunction of these factors leads to a long list of realizable embedded systems, making it challenging to design a general emulator for firmware re-hosting.

These challenges have encouraged the development of a variety of emulation solutions. A widely adopted approach is the *hardware-in-the-loop* (HITL) [9–12], in which the firmware is only partially emulated and whenever an unsupported I/O operation is attempted, the request is redirected to the hardware itself. In HITL, the firmware interacts with a hardware platform that mimics the real hardware. The platform might be an actual piece of hardware or a hardware simulator and provides the necessary peripheral interfaces and inputs/outputs that the firmware requires to function correctly. This lessens the need for access to the real target hardware and enables testing and evaluation of the firmware in a controlled environment. Various studies employ operating system and/or hardware abstractions [13–15] to take advantage of the abstraction layer provided by the firmware. An operating system abstraction is a layer of software that provides a standard interface for the firmware to interact with the hardware, while a hardware abstraction provides a similar layer of abstraction, but it is specific to the hardware being used. Such abstractions provide high-level representations of the underlying hardware, thus enabling the firmware to interact with the hardware in a manner that is independent of the actual hardware's implementation. Conversely, other studies aim for full-system emulation [16–18] without the presence of actual hardware. With full-system emulation, the behavior of an entire embedded system, including the firmware and the underlying hardware, is recreated in a virtual environment. These works focus on the automated creation of models that describe the interactions between firmware and hardware, allowing for the replaying of these interactions without a direct connection to the specific device or even the learning of these interactions through models generated from recorded real interactions. Once the software and hardware models have been created, they can be integrated into a virtual environment that simulates the behavior of the real system. The virtual environment can be used to test and evaluate the firmware in a manner that is independent of the underlying hardware. This makes it possible to test and evaluate firmware on different hardware platforms, or even on platforms that do not yet exist, without the need for access to the actual hardware.

Although much improvement has been made to handle the firmware re-hosting challenge, the current approaches come with limitations. HITL-based solutions are effective in enabling interactivity and allowing testers to utilize dynamic analysis tools to input data into the firmware. However, this method introduces latency in the forwarding process, thus impeding the execution speed, reducing parallelism and scalability, and limiting its performance as a testing approach. Furthermore, HITL still entails a substantial tie between the firmware and hardware. Methods that rely on operating system or hardware abstractions exceed the HITL drawbacks, though these approaches have limitations in terms of the types of firmware they can handle. Indeed, in order to accommodate a broad spectrum of firmware, it is crucial for emulators to be devoid of high-level abstractions. Finally, learning-based solutions still require interactions with real hardware to collect data on the peripherals' behavior. To overcome the limitations of the currently adopted methodologies for re-hosting, we propose a novel approach that takes advantage of binary rewriting techniques. Binary rewriting is meant to modify the behavior of a compiled program without having its source code or recompiling it while keeping the binary executable. Binary rewriting can be classified as either static or dynamic. Static binary rewriting involves making modifications to the binary file and saving them permanently, while dynamic binary rewriting modifies the binary while it's being executed, without making any permanent change.

Several binary rewriting methodologies have been proposed in literature, including both static [19–23] and dynamic [24–28] solutions. Binary rewriting can be applied in a variety of ways, such as monitoring a program during execution, optimizing it, and emulating it [29].

Finally, many proposals build the emulation on top of base emulators such as QEMU [30] and Simics [31]. The exclusive use of these tools for hardware emulation is indeed discouraged since they are only able to emulate a restricted list of CPUs and peripherals and support a limited number of possible configurations [30] and may also require significant analyst intervention [31].

4 State-of-the-Art Approaches and Their Limitations

Firmware re-hosting has drawn a lot of interest in the literature, and various works came up introducing solutions to address the hardware emulation challenge with the ultimate goal of enabling faster security assessment of firmware.

HITL-Based Approaches
Several works such as Avatar [9], Avatar2 [32], PROSPECT [10], SURROGATES [11], Inception [33], and Charm [12] pursue the HITL approach, proposing partially emulation of firmware with unsupported I/O requests redirected to real peripherals, which still implies a strong dependence from the hardware. These works propose, in order to conduct dynamic analysis, to partially offload the execution of firmware to actual hardware, thus compromising scalability.

Abstraction-Based Approaches
Other works design emulation on top of OS abstractions [13, 14] or hardware abstraction layers (HALs) [15]. Firmadyne [13] is a full-system emulation tool for automated large-scale dynamic analysis of firmware. When a firmware image is provided to Firmadyne, the tool extracts the file system and performs an analysis to determine the hardware specifics. Then, a pre-built Linux kernel that corresponds to these specifics is employed, and an initial emulation is conducted using QEMU to infer the system and network configuration. Costin et al. [14] present a framework for scalable security testing of embedded web interfaces using dynamic analysis tools. The framework relies on the emulation of firmware images via QEMU, by replacing the system native kernel with a default kernel – for a specific CPU architecture – supported by QEMU. The limit of relying on OS abstraction is that only a reduced number of firmware are supported by the framework; indeed only Linux-based firmware is handled by both [13, 14]. HALucinator [15] relies on a technique called high-level emulation (HLE) to perform dynamic analysis on firmware in embedded systems. The authors leverage the use of hardware abstraction layers (HALs) commonly used by firmware developers to simplify their jobs, as a basis for re-hosting and analyzing firmware. The technique works by first identifying the library functions responsible for hardware interactions in a firmware image and then providing high-level replacements for these functions in a full-system emulator such as QEMU. The authors demonstrated the practicality of HLE for security analysis by supplementing their prototype system, HALucinator, with a fuzzer to locate multiple previously unknown vulnerabilities in firmware middleware libraries. In [34] a technique called para-rehosting to make re-hosting of microcontroller (MCU) software to commodity hardware smoother is proposed. The authors implemented a portable MCU (PMCU) using the POSIX interface, which models common functions of the MCU cores and accurately replicates the common behaviors of real MCUs. They abstracted and modeled common functions of MCU cores and proposed HAL-based peripheral function replacement, in which high-level hardware functions are replaced with an equivalent back-end driver on the host, allowing for incremental plug-and-play library porting. Both [15] and [34] present

interactive and hardware-independent environments. However, these environments are built over the assumption that the firmware relies on HALs, which may not always be the case. In order to be able to deal with a wider range of firmware, emulators should be abstraction-free, meaning that they should not depend on high-level constructs.

Learning-Based Approaches
Other approaches have been proposed to automatically create emulators for firmware, which require knowledge of how the firmware interacts with the peripherals, by capturing and reproducing data generated during I/O interactions to model the hardware behavior. This allows large-scale and interactive executions, but inevitably necessitates trace recording from within the device itself, thereby restricting accurate execution in the emulator to only the recorded program paths. In [16–18, 35, 36] the peripherals' behavior is learned and the interactions between the firmware and hardware are modeled in order to enable virtualized execution of firmware without implement peripheral emulators at all. Pretender [16] and Conware [35] gather a set of observations of the low-level interactions between the firmware and the original peripherals by means of HITL or code instrumentation. Pretender then utilizes machine learning to generate models of the memory-mapped input/output (MMIO) operations and interrupt-driven peripherals, which can replace the physical hardware. In contrast, Conware generates composable automata representations based on the collected recordings to model the peripherals, which can be merged to build generalized models. Both Pretender and Conware intend to emulate arbitrary firmware without having to instrument the actual firmware; however, they imply an access to the physical hardware during the training phase in order to gather the observations needed for the emulation and require instrumentation to detect interactions with the hardware. Also, Pretender only provides models for interrupt-based firmware that therefore are not generic and suitable for non-interrupt-based firmware. P^2IM [17] is a framework for MCU firmware *approximate* emulation that does not provide model for the physical peripherals themselves, but treats them as black boxes. Whenever the running firmware requests interactions with the peripherals, it is provided with acceptable inputs that simply satisfy internal checks and do not cause the execution to halt or crash. P^2IM does not require a deep knowledge about the peripherals behavior, allowing only a small set of values to be considered as possible inputs to the firmware. Although this functioning allows hardware-independent emulation, it reduces the ability to represent effectively complex firmware logic.

Symbex-Based Approaches
Other works achieve firmware re-hosting by means of symbolic execution [18, 36, 37]. Laelaps [18] and Jetset [36] propose a symbolic execution-based approach to infer the behavior of the peripheral expected by the firmware. Laelaps is a concolic execution-based firmware re-hosting framework that combines concrete and symbolic execution. It uses a full-system emulator such as QEMU to run the firmware and gain the inner state of the execution and switches to symbolic execution whenever an access to an unimplemented peripheral is attempted, in

Table 1 Strengths and limitations of existing approaches

Approaches	Articles	Strengths	Limitations
HITL	[9], [32], [10], [11], [33], [12]	– Interactivity; – Dynamic analysis enabled	– Hardware dependency – Limited scalability – Latency
Abstraction-based	[13], [14], [15]	– Hardware independence	– Abstractions not always available
Learning-based	[16–18, 35, 36]	– Hardware independence after training	– Hardware dependency during dataset recording
Symbex-based	[18, 36, 37]	– Extensive path exploration – Possibility to automate test case generation	– Potential path explosion – Problematic complex behavior modeling

order to find valid input that leads the execution to a path that resembles a realistic behavior. In order to prevent the path explosion, Laelaps relies on the Context Preserving Scanning Algorithm (CPSA) heuristics, which are able to infer inputs valid for the near future execution, but which may cause execution crash in the long term. Jetset is a tool that relies on symbolic execution for inferring how the peripheral devices are expected to behave in the interaction with the firmware. These inferred are used while the firmware is running into an emulator such as QEMU in order to reproduce a device target functionality. Path explosion is mitigated using guided symbolic execution with a variation of Tabu Search to minimize the distance to the goal. However, following this approach, at each branch, the direction to take is chosen by looking at the distance to the goal, which can make it difficult to model more complex behaviors. μEmu [37] uses symbolic execution to extract valuable information and to build a knowledge base that is used to emulate the peripherals behavior during firmware re-hosting. As it carries the knowledge extraction process, μEEmu tries to avoid the path explosion by switching to another path only when the current one is found invalid. However, it fails to emulate complex peripherals' behaviors (Table 1).

5 Our Approach

We propose an approach to enable dynamic security assessment techniques such as fuzzing on firmware which extends and somehow redefines our proposal in [38]. Our proposal relies on binary rewriting to obtain a full-system emulation of firmware, ensuring hardware independence as well as interactivity and overcoming the limitations experienced in the current approaches. By providing innovative insights and practical solutions to the security concerns surrounding IoT devices, this chapter contributes directly to the advancement of knowledge and practices within the realm of the device edge-cloud continuum. Our research underscores the significance of addressing vulnerabilities in IoT devices and highlights the potential for improved paradigms, architectures, and applications in ensuring the integrity and resilience of this interconnected ecosystem. Besides the drawbacks already discussed in Sect. 3, most of the current approaches involve the use of binary instrumentation to intercept the invocations to functions related to I/O interactions with peripherals and forward them to their replacement models. Since the instrumented code is meant to be executed in a virtual environment outside the firmware execution environment, its use is significantly slowing down the process.

In our proposal, we completely bypass that step by fully replacing the interactions between the firmware and the hardware with code. We integrate the embedded peripherals' behavior at a high level by firmware rewriting, avoiding the involvement of lower-level abstractions such relying on OS assumptions or using HALs. In this way, we enable the application of vulnerability assessment techniques based on a large number of executions and possibly crashes of the binaries under analysis.

The process involves the following steps: (i) Portions of the binary code that constitute interactions with the peripherals must be identified. This step can be accomplished through various methods such as manual reverse engineering, debugging firmware in a hardware-in-the-loop environment, or locating HAL functions via library matching. (ii)Once these functions are identified, their behavior must be rewritten to ensure successful emulation of the firmware, which is a challenging task that relies on manual development due to its difficulty to automate. However, literature suggests that several approaches can be used to automatically develop models to replace the original functions by recording firmware interactions with hardware peripherals, as discussed in Sect. 4. (iii)To perform the actual replacement of functions that interact with hardware with ad hoc implementations, we rely on binary rewriting techniques. This is the most significant aspect of our proposal as it avoids intercepting calls using binary instrumentation approaches, thereby significantly speeding up the emulation process. The binary rewriting step can be performed in several ways, some of which have already been introduced in Sect. 4.

The most significant benefit of rewriting, compared to current approaches, is that the firmware can be treated as normal software from an emulation perspective. As a result, the entire re-hosting process is much faster, and vulnerability assessment techniques such as fuzzing can be easily adopted without incurring excessive slowdown due to binary instrumentation.

In fact, what happens during the normal operation of a device is that numerous interactions occur with the hardware. By choosing to use binary instrumentation techniques to identify these interactions, the peripheral emulation process undergoes a significant slowdown.

For example, consider an industrial control system (ICS) that leverages REST APIs to expose the status of a humidity sensor. In particular, when a request for humidity data is held through such APIs, the ICS calls a library function supplied by the manufacturer, which in turn invokes another function meant to interact with the peripherals, which is an HAL function provided by the microcontroller vendors,

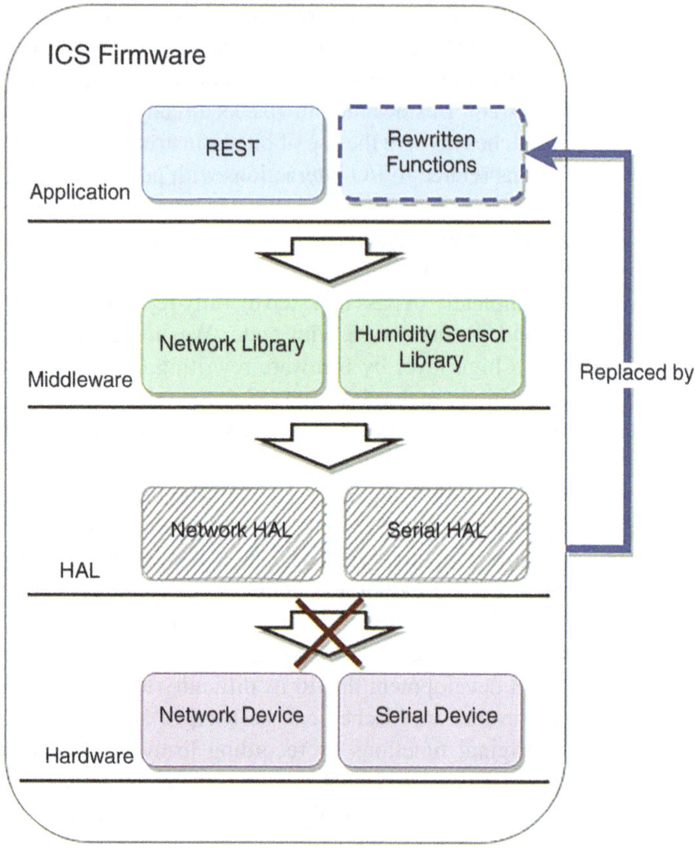

Fig. 1 Example: ICS rewritten firmware

able to read the humidity value from the chip, by means of a serial communication. It is well known that the REST service relies on the HTTP protocol, which is implemented on top of a TCP communication that in our example is put on in a library using an HAL to communicate with the Ethernet port. Even considering such a simple scenario, we can identify several interactions between firmware and hardware, e.g., function calls to the microcontroller HAL and communication with the Ethernet port. Our idea consists in replacing these functions with custom implementations, in order to be able to dissolve the firmware-hardware bonds and achieve complete independence from the underlying hardware, as illustrated in Fig. 1. By rewriting the functions related to I/O interaction, we are able to achieve firmware emulation without the need of instrumenting the emulation.

6 Conclusions and Future Work

We inspected the state-of-the-art of firmware re-hosting for vulnerability assessment purpose, and we analyzed extensively the strengths and weaknesses of existing solutions. In particular, we have analyzed how the use of dynamic analysis techniques that require a high number of executions to be performed quickly as occurs in fuzzing turns out to be impractical. Our research offers an alternative solution that utilizes binary rewriting to address the firmware re-hosting challenges, thereby enabling firmware to be tested using dynamic analysis techniques more efficiently. The chapter aims to initiate a dialogue in the field of rapid firmware fuzzing and posits that the proposed methodology can enhance the existing vulnerability assessment methodologies of analysts. Currently, we are in the process of implementing the proposed approach and identifying the optimal techniques to employ at each stage of the rewriting process. Our preliminary findings are promising and show the effectiveness of our proposal.

References

1. G. Fortino, A. Guzzo, M. Ianni, F. Leotta, M. Mecella, Exploiting marked temporal point processes for predicting activities of daily living, in *2020 IEEE International Conference on Human-Machine Systems (ICHMS)* (IEEE, 2020), pp. 1–6
2. G. Fortino, A. Guzzo, M. Ianni, F. Leotta, M. Mecella, Predicting activities of daily living via temporal point processes: approaches and experimental results. Comput. Electr. Eng. **96**, 107567 (2021)
3. G. Fortino, A. Guerrieri, P. Pace, C. Savaglio, G. Spezzano, IoT platforms and security: an analysis of the leading industrial/commercial solutions. Sensors **22**(6), 2196 (2022)
4. Y. He, Z. Zou, K. Sun, Z. Liu, K. Xu, Q. Wang, C. Shen, Z. Wang, Q. Li, {RapidPatch}: firmware hotpatching for {Real-Time} embedded devices, in *31st USENIX Security Symposium (USENIX Security 22)* (2022), pp. 2225–2242
5. A. Guzzo, M. Ianni, A. Pugliese, D. Saccà, Modeling and efficiently detecting security-critical sequences of actions. Futur. Gener. Comput. Syst. **113**, 196–206 (2020)
6. M. Salehi, L. Degani, M. Roveri, D. Hughes, B. Crispo, Discovery and identification of memory corruption vulnerabilities on bare-metal embedded devices. IEEE Trans. Dependable Secure Comput. **20**(2), 1124–1138 (2022)
7. [Online]. Available: https://lcamtuf.coredump.cx/afl/
8. Y. Shoshitaishvili, R. Wang, C. Salls, N. Stephens, M. Polino, A. Dutcher, J. Grosen, S. Feng, C. Hauser, C. Kruegel, G. Vigna, SoK: (state of) the art of war: offensive techniques in binary analysis, in *IEEE Symposium on Security and Privacy* (2016)
9. J. Zaddach, L. Bruno, A. Francillon, D. Balzarotti et al., Avatar: a framework to support dynamic security analysis of embedded systems' firmwares, in *NDSS*, vol. 14 (2014), pp. 1–16
10. M. Kammerstetter, C. Platzer, W. Kastner, Prospect: peripheral proxying supported embedded code testing, in *Proceedings of the 9th ACM Symposium on Information, Computer and Communications Security* (2014), pp.329–340
11. K. Koscher, T. Kohno, D. Molnar, {SURROGATES}: enabling {Near-Real-Time} dynamic analyses of embedded systems, in *9th USENIX Workshop on Offensive Technologies (WOOT 15)* (2015)

12. S.M.S. Talebi, H. Tavakoli, H. Zhang, Z. Zhang, A.A. Sani, Z. Qian, Charm: facilitating dynamic analysis of device drivers of mobile systems, in *27th USENIX Security Symposium (USENIX Security 18)* (2018), pp. 291–307
13. D.D. Chen, M. Woo, D. Brumley, M. Egele, Towards automated dynamic analysis for linux-based embedded firmware, in *NDSS*, vol. 1 (2016), pp. 1–1
14. A. Costin, A. Zarras, A. Francillon, Automated dynamic firmware analysis at scale: a case study on embedded web interfaces, in *Proceedings of the 11th ACM on Asia Conference on Computer and Communications Security* (2016), pp. 437–448
15. A.A. Clements, E. Gustafson, T. Scharnowski, P. Grosen, D. Fritz, C. Kruegel, G. Vigna, S. Bagchi, M. Payer, {HALucinator}: firmware re-hosting through abstraction layer emulation, in *29th USENIX Security Symposium (USENIX Security 20)* (2020), pp. 1201–1218
16. E. Gustafson, M. Muench, C. Spensky, N. Redini, A. Machiry, Y. Fratantonio, D. Balzarotti, A. Francillon, Y.R. Choe, C. Kruegel et al., Toward the analysis of embedded firmware through automated re-hosting, in *22nd International Symposium on Research in Attacks, Intrusions and Defenses (RAID 2019)* (2019), pp. 135–150
17. B. Feng, A. Mera, L. Lu, {P2IM}: scalable and hardware-independent firmware testing via automatic peripheral interface modeling, in *29th USENIX Security Symposium (USENIX Security 20)* (2020), pp. 1237–1254
18. C. Cao, L. Guan, J. Ming, P. Liu, Device-agnostic firmware execution is possible: a concolic execution approach for peripheral emulation, in *Annual Computer Security Applications Conference* (2020), pp. 746–759
19. E. Bauman, Z. Lin, K.W. Hamlen et al., Superset disassembly: statically rewriting x86 binaries without heuristics, in *NDSS* (2018)
20. J.R. Larus, T. Ball, Rewriting executable files to measure program behavior. Softw.: Pract. Experience **24**(2), 197–218 (1994)
21. G. Ravipati, A.R. Bernat, N. Rosenblum, B.P. Miller, J.K. Hollingsworth, Toward the deconstruction of dyninst. University of Wisconsin, Technical report, vol. 32, 2007
22. D.W. Wall, Systems for late code modification, in *Code Generation–Concepts, Tools, Techniques: Proceedings of the International Workshop on Code Generation* (Springer, London, 1992), pp. 275–293
23. L. Van Put, D. Chanet, B. De Bus, B. De Sutter, K. De Bosschere, Diablo: a reliable, retargetable and extensible link-time rewriting framework, in *Proceedings of the Fifth IEEE International Symposium on Signal Processing and Information Technology, 2005* (IEEE, 2005), pp. 7–12
24. K. Scott, J. Davidson, Strata: a software dynamic translation infrastructure, in *IEEE Workshop on Binary Translation* (2001)
25. C. Cifuentes, B. Lewis, D. Ung, Walkabout-a retargetable dynamic binary translation framework, in *Workshop on Binary Translation* (2002), pp. 22–25
26. J.K. Hollingsworth, B.P. Miller, J. Cargille, Dynamic program instrumentation for scalable performance tools, in *Proceedings of IEEE Scalable High Performance Computing Conference* (IEEE, 1994), pp. 841–850
27. B. Buck, J.K. Hollingsworth, An API for runtime code patching. Int. J. High Perform. Comput. Appl. **14**(4), 317–329 (2000)
28. C.-K. Luk, R. Cohn, R. Muth, H. Patil, A. Klauser, G. Lowney, S. Wallace, V.J. Reddi, K. Hazelwood, Pin: building customized program analysis tools with dynamic instrumentation. ACM SIGPLAN Not. **40**(6), 190–200 (2005)
29. M. Wenzl, G. Merzdovnik, J. Ullrich, E. Weippl, From hack to elaborate technique–a survey on binary rewriting. ACM Comput. Surv. (CSUR) **52**(3), 1–37 (2019)
30. F. Bellard, QEMU, a fast and portable dynamic translator. in *USENIX Annual Technical Conference, FREENIX Track*, vol. 41, no. 46. Califor-nia, USA (2005), pp. 10–5555
31. P.S. Magnusson, M. Christensson, J. Eskilson, D. Forsgren, G. Hallberg, J. Hogberg, F. Larsson, A. Moestedt, B. Werner, Simics: a full system simulation platform. Computer **35**(2), 50–58 (2002)

32. M. Muench, D. Nisi, A. Francillon, D. Balzarotti, Avatar 2: a multi-target orchestration platform, in *Proceedings Workshop Binary Analysis Research (Colocated NDSS Symposium)*, vol. 18 (2018), pp. 1–11
33. N. Corteggiani, G. Camurati, A. Francillon, Inception: {System-Wide} security testing of {Real-World} embedded systems software, in *27th USENIX Security Symposium (USENIX Security 18)* (2018), pp. 309–326
34. W. Li, L. Guan, J. Lin, J. Shi, F. Li, From library portability to para-rehosting: natively executing microcontroller software on commodity hardware (2021). arXiv preprint arXiv:2107.12867
35. C. Spensky, A. Machiry, N. Redini, C. Unger, G. Foster, E. Blasband, H. Okhravi, C. Kruegel, G. Vigna, Conware: automated modeling of hardware peripherals, in *Proceedings of the 2021 ACM Asia Conference on Computer and Communications Security* (2021), pp. 95–109
36. E. Johnson, M. Bland, Y. Zhu, J. Mason, S. Checkoway, S. Savage, K. Levchenko, Jetset: targeted firmware rehosting for embedded systems, in *30th USENIX Security Symposium (USENIX Security 21)* (2021), pp. 321–338
37. W. Zhou, L. Guan, P. Liu, Y. Zhang, Automatic firmware emulation through invalidity-guided knowledge inference, in *USENIX Security Symposium* (2021), pp. 2007–2024
38. G. Fortino, C. Greco, A. Guzzo, M. Ianni, Enabling faster security assessment of re-hosted firmware, in *2022 IEEE International Conference on Dependable, Autonomic and Secure Computing, International Conference on Pervasive Intelligence and Computing, International Conference on Cloud and Big Data Computing, International Conference on Cyber Science and Technology Congress (DASC/PiCom/CBDCom/CyberSciTech)* (IEEE, 2022), pp. 1–6

Performance Analysis of a Blockchain for a Traceability System Based on the IoT Sensor Units Along the Agri-Food Supply Chain

Maria Teresa Gaudio, Sudip Chakraborty, and Stefano Curcio

1 Introduction

The agri-food supply chain can be seen as a complex system of systems (SOS) [1], and a traceability system along the entire supply chain seems challenging to realize.

Today, different technologies exist to ensure traceability [2–8], but at the same time, some critical points remain and influence the reliability of the entire system.

In particular, each different product follows a specific supply chain with its requirements and constraints. Most of the critical points are represented by the interactions between different actors involved in the supply chain, where the risk can increase due to the lower automated layers of protection and thus, malicious [9] – people or others – could intervene, causing fraud and damage to the final product and the entire supply chain.

The insertion of the IoT sensor unit in correspondence with these critical points could be interesting concerning the real-time monitoring and consequent checking of the process and at the same time, the possible integration with blockchain technology. This last perfectly responds to the four steps of a traceability system – identification, recording, data links, and report [10, 11] – with its fundamental principles: immutability and transparency, disintermediation and provenance, and trust and agreement [12]. Moreover, blockchain technology could represent a solution that can be implemented in all specific supply chains [13, 14].

This works referred to a multilayered solution for the agri-food supply chain traceability proposed in [15]. In this paper, a more in-depth description of the blockchain setup technology in the Hyperledger Fabric environment was described, and the main results of transaction simulation were presented. Compared to existing

M. T. Gaudio (✉) · S. Chakraborty · S. Curcio
Università della Calabria, Rende, Italy
e-mail: mariateresa.gaudio@unical.it; sudip.chakraborty@unical.it; stefano.curcio@unical.it

C. Savaglio et al. (eds.), *Device-Edge-Cloud Continuum*, Internet of Things,
https://doi.org/10.1007/978-3-031-42194-5_3

technologies, the multilayer proposed solution is referred to as the whole agri-food supply chain system, and not to a single stage of the specific supply chain.

2 Step-by-Step Hyperledger Blockchain Setting

To consider a generic scheme solution for all types of supply chains, the entire agri-food supply chain was considered with only three main different actors involved: farmer, manufacturer, and distributor. The manufacturer considers the producing and packaging phase as the only one, and the waste management unit is neglected, considering the real hypothesis of zero waste in the supply chain.

Using Hyperledger Fabric, these hypotheses are more readily accepted, because Hyperledger technology is the most scalable among existing blockchain solutions.

Therefore, when it will be necessary for the next, the adding of other organizations will be possible and easy to adopt, if the hyperledger initial setup will be performed correctly. Moreover, hyperledger can be set as public or private, both permissioned and permissionless; i.e., it provides very high modularity, capable of adapting to any need [16].

Unlike other blockchain platforms, hyperledger is an open source hosted by Linux Foundation. The main problem could be the high memory capacity that the machine in use must have to be able to memorize the blocks of the chain, to which new information will be added each time. For this reason, it was decided to start with the installation of the blockchain solution on the virtual machines (VMs), in order to preserve the integrity of the machine and at the same time, use a cloud solution to store all data and blocks. Obviously, also this choice will involve a cost, but at the same time, this could be managed in the best possible way from time to time, without going to wear out the machine. This paragraph shows the step-by-step working procedure for the hyperledger blockchain solution used.

2.1 General Architecture

The general architecture proposed for the agri-food supply chain traceability consists of the management of information from farm to fork, i.e., from the agricultural phase to the end consumer.

The tracking and reconstruction of the information flow passes through the interactions between the different actors involved in the entire supply chain.

For this general application, the blockchain chosen is an open-permissioned Hyperledger Fabric blockchain. And starting from this case study, extra-virgin olive oil was considered by four organizations:

- Organization 1 (Org1) involves the interactions between the manufacturing process and the agricultural phase, where the raw material comes from.

- Organization 2 (Org2) involves the interactions between the final product coming from the manufacturing process and the distribution phase.
- Organization 3 (Org3) refers to the interactions involved in the product recognition, e.g., in the store or a point of consumption.
- Organization 4 (Org4) is the orderer organization.

Each organization has two peers, while the orderer organization has three orderer peers.

The ordering service is one of the main features of Hyperledger Fabric. It guarantees transaction ordering. In fact, Fabric relies on deterministic – not probabilistic – consensus algorithms; thus, any block validated by the peer is guaranteed to be final and correct.

Moreover, separating the endorsement of chaincode execution (which happens at the peers) from ordering gives Fabric advantages in performance and scalability, eliminating bottlenecks which can occur when execution and ordering are performed by the same nodes.

In brief, this project has three organizations defined which can contain the information – regarding farming, manufacturing, and distribution – a single orderer and a single channel for this business network.

The entities will interact with the blockchain application by invoking chaincode in the fabric network, updating the ledger world state, and writing transaction logs. The blockchain network was set up through the use of 4 Google VMs. In the creation of VMs, it is convenient to identify the geographical area that allows us to spend less.

Among all platforms, Google was chosen thinking about the possibility by a company already has a Google account, and thus, there is no need to open another accounting service on another platform. Moreover, Google already returns some performance evaluations inherent to the machine.

As an operating system, Ubuntu 18.04 LTS was installed on each VM. All VMs have two vCPU and 4 GB of memory. In contrast, VM1 has 8 GB because it is the one that will store more data from different operations: the network configuration, the creation of the crypto-material for each organization, the creation of the channel artifacts, and the creation of a docker swarm network.

In this network, all ORGs can interact in one channel (see Fig. 1). The channel was generically called "mychannel." Docker Compose was used to launch the corresponding fabric containers, and as the first step, the services run in the container are defined in a Docker Compose YAML file.

The installation of Hyperledger Fabric was carried out according to the indications of Hyperledger Fabric manual version 2.3 [17], which already contains all packages' installation about prerequisites, among valuable docker for managing the containers.

The first thing to do is the creation of the artifacts folder for the channel configuration.

In a config folder, different configuration YAML files are present, and they are used to track what kind of flow of information is necessary to follow, i.e., which chaincode deployment you want to perform. To make it, Membership Service Provider (MSP) is necessary for each organization.

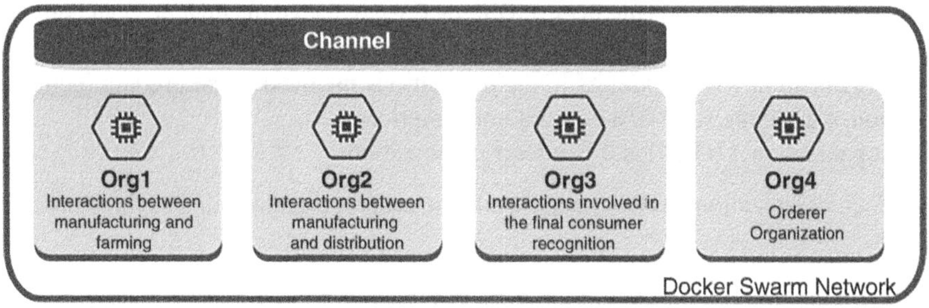

Fig. 1 Hyperledger Fabric architecture for a generic agri-food supply chain

First of all, let us start with the creation of the crypto-materials for each organization. Thanks to the Docker Compose YAML file and a shell script file (.sh) – written in a bash language and executable directly in the terminal – the certificate authorities (CAs), MSP, and Transport Layer Security (TLS) certificates are generated in a special crypto-config folder for each organization.

After the certificate's creation, it is time to create Genesis block and channel transaction files. Through another shell script file, named "create-artifacts," the Genesis block, Genesis channel configuration, and Anchor Peer transactions are generated. This file configures a single anchor peer for each organization.

The anchor peer is more critical in Fabric, because it is a peer node that enables the communication between peers of different organizations and discovers all active channel participants. At this point, it is essential to share all information created among all different organizations, through the creation of a particular repository on GitHub, or uploading each organization's certificates to the others or again, creating at the starting point only a first VM and then, cloning its disk and creating from it the other VMs.

The creation of a docker swarm network is necessary to orchestrate each instruction in each VM and therefore, in each organization.

After the docker swarm network creation, all containers have to run on each organization. It was made through another appropriate Docker Compose YAML file. In this Docker Compose file, each peer has a CouchDB database, where the information is collected.

2.2 *Smart Contract Operations*

In Hyperledger Fabric, the smart contract is called "chaincode," and for this case study, a specific chaincode, called "Foodchain," was written in the Go language. Fabric also supports other programming languages, but it was demonstrated that Go is the most powerful for this technology [18].

Table 1 General foodchain structure algorithm in natural language

Algorithm 1 – Foodchain		
1:	INITIALIZE	the executable function through the main package
2:	IMPORT	all necessary Go packages to execute different functions
3:	DEFINE	the FoodchainContract function to manage the Asset, i.e., the information coming from IoT Sensor Units
4:	DEFINE	user-defined type structures to store a collection of the product and the participants
5:	INITIALIZE	of FoodchainContract
6:	INVOKE	createProduct function to create the block and to add information for a certain product
7:	INVOKE	manufactureProcessing function to add information about the manufacturing process
8:	INVOKE	distributorProcessing function to update information about the distribution state
9:	INVOKE	query function about reading the status of the product at a given time
10:	IF ELSE RETURN	query function reads a consistent status of the product at a given time, it means that information and condition in input to the blockchain are correct, error
11:	INVOKE	setupFoodchainTracer function to record all information in the same block
12:	RETURN	all information contained in a block

Go is a perfect programming language to develop fast and scalable blockchain systems. Go language is not only simple to learn; it also comes with the best features of JavaScript and Python, such as user-friendliness, scalability, stability, and speed – anything that makes it the best choice to offer tailored-made blockchain applications.

The Foodchain chaincode is reported and described in natural language, cf. Table 1.

The generical algorithm of Foodchain also contains other small functions, which can be written in the same Foodchain or also in the subsequent shell script files for the working of Hyperledger, that is, during the initializing, invoking, and committing of chaincode in different involved organizations, valuable steps to execute the transaction of the asset information.

In these shell script files, when a function is invoked on peers, replacing the localhost with the VM's IP address corresponding to the orderer organization is essential. This replacement is necessary to enable the functionality of the ordering service among the other organizations.

In each of these shell script files, before each function, the set of environmental variables and the declaration of paths for each peer of each corresponding organi-

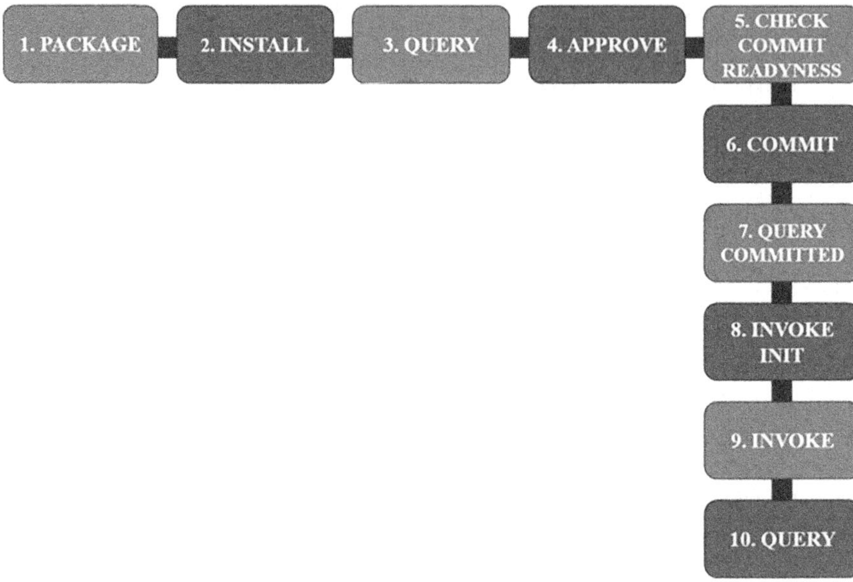

Fig. 2 Chaincode life cycle

zation were defined and assigned. Through these files, each organization can create and join the channel.

Foodchain have to undergo a series of operational steps, so that transactions on the blockchain network can be carried out. These operations identify the chaincode life cycle (see Fig. 2) and they carry out from other executable shell script files.

In the next paragraph, some of their life cycle operations were printed and shown, within other significant results of the transaction simulations. To avoid errors, each operation must be executed in each organization, in its corresponding VM.

The same primary performance results were obtained using the Hyperledger Caliper tool [19]. It is a blockchain benchmark tool that allows users to measure the performance of a blockchain implementation, starting by using predefined use cases. Caliper quickly returns the number of successful and failed operations, the send rate and throughput – expressed in transactions per second (TPS) – and different values of latency, respectively, minimum, maximum, and average. In this work, these values were used to elaborate more complete results beneficial to describe the comprehensive performance of the blockchain solution proposed.

3 Results of Simulations

As mentioned before, the first results shown in the following figures regard the chaincode life cycle on VM1, which correspond to the Org1 (see Figs. 3, 4

Fig. 3 Chaincode installation on VM1

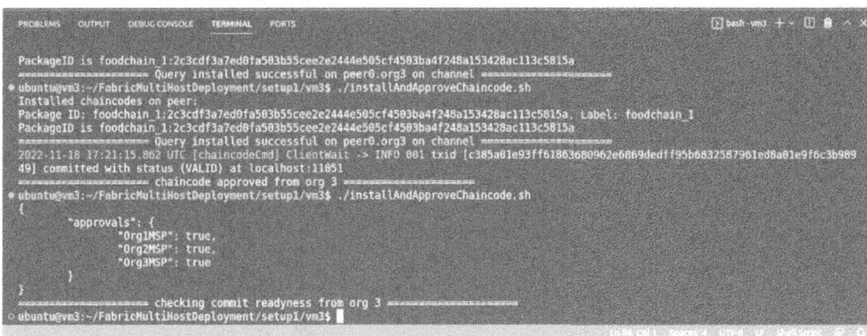

Fig. 4 Chaincode approval on VM1

and 5) useful to execute transactions. These last figures show the screens with the instructions passed to the terminal for the chaincode deployment in VM1 and their respective results respectively for the installation, the approval, and the commitment of the chaincode.

Other interesting results could be observed in terms of the execution time, average latency, throughput, and scalability.

Considering the nature of this blockchain, where the number of organizations has been minimized, the performance of a single peer was evaluated.

The execution time generally increases with the increase of number of transactions. Figure 6 shows a comparison between the execution time for query and invoke operations was done. The execution time seems to be comparable for query and invoke functions. Both tend to increase at a higher number of transactions. At 100 transactions, the execution time for invoke is equal to 58 seconds, while for query it is 50 seconds.

The transaction latency is the time between the sending request and the receiving of the response by the client. In Hyperledger Fabric, two different types of latency exist: query latency and update latency.

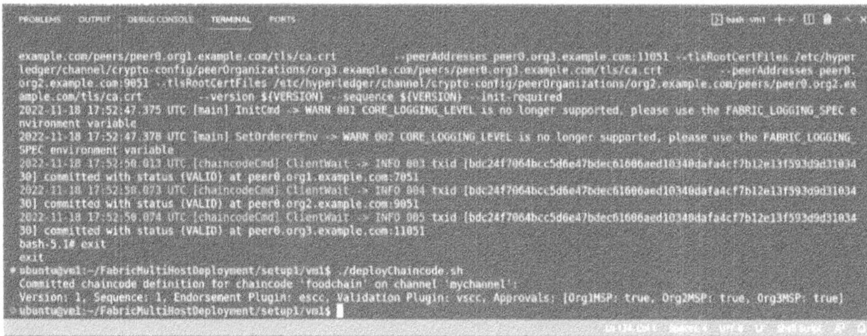

Fig. 5 Chaincode committed on VM1

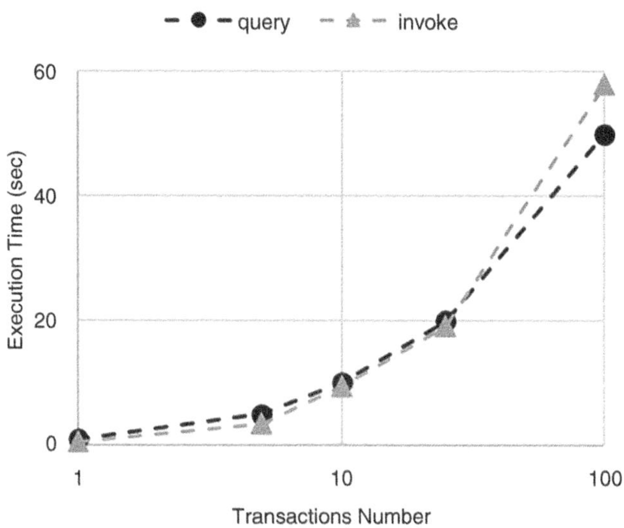

Fig. 6 Execution time

The first is the time between the request sent and the response received by the client. The second is the time between the request sent and the confirmation event – which notified that the transaction occurred and entered in a block, adding it to the blockchain network – is received by the client.

Considering the entire flow of information, query and update latencies – respectively, *LpQ* and *LpT* – can be expressed as follows:

$$LpQ = Tapp + Tendorser + Tcc + Tconcurrent\_transaction+ \\ + Tworldstate \quad (1)$$

$$LpT = Tapp + Tendorser + Tcc + Tconcurrent\_transaction + \\ + Tworldstate + Torderer + Tpeer \quad (2)$$

where:

- *Tendorser* is the time it takes for a transaction to reach enough endorser peers to satisfy the policy and send a response. This time is affected by the number of endorser peers of the blockchain network and also influenced by the network latency and the quality of hardware on which the endorser peers are running.
- *Tcc* is the overall execution time of a chaincode and it also depends on the endorsement policy.
- *Torderer* is when the ordering service receives transactions and collects them in the same block. This time depends on the ordering service used. It may be implemented in one mode, i.e., with only one node involved, or with Kafka mode, where there is also coordination among nodes that constitute the ordering. In this case, the network latency impacts reaching the orderer where it is running. Moreover, the block size and the timeout parameters – that is, the maximum helpful time to create a block – influence latency. In other words, if a block is more significant or has a higher timeout, more time waits for the orderer to create the block, consequently increasing latency.
- *Tpeer* is the total time required by peers to receive a transaction block, and validate each of them one by one, to append the block to its ledger. This time depends on the number of peers present in the network, the network latency, and the hardware capacity of each peer, as they have to validate all transactions of a block.
- *Tconcurrent_transaction* refers to the number of concurrent transactions that are executed; more concurrent transactions occur, more endorsers and peers are charged and thus, the average latency increases.
- *Tworldstate* is the time spent in reading and writing operations on the worldstate and it depends on the type of database used. In this case study, CouchDB was used.

Comparing Eqs. 1 and 2, the difference between update latency and query latency corresponds to the sum of Tpeer and Torderer.

In Fig. 7, these two latencies are compared, and the difference between them represents just *Tpeer* + *Torderer*. It can be seen that the update latency is almost twice than query latency and *Tpeer* + *Torderer* is equal to 8.4 seconds for 100 transactions.

The throughput of transactions is the rate at which valid transactions are committed by the blockchain network in a defined period. This rate is not for a single node, but refers to an entire network system. It is expressed as TPS. Figure 8 compares the throughput of the query and invoke functions. Considering latency is higher for invoking the function, the query function seems to have a throughput higher than invoking the function.

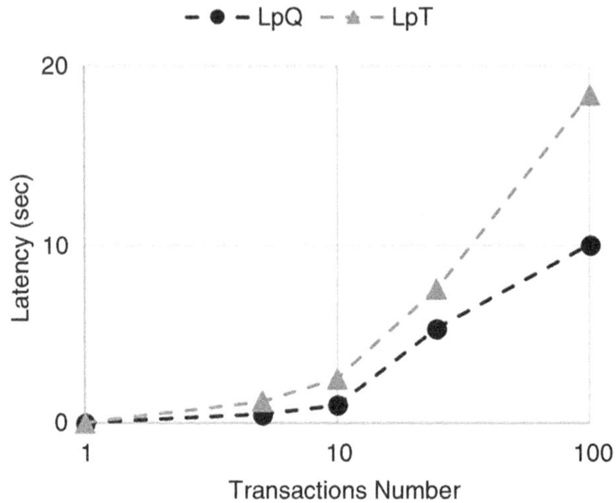

Fig. 7 Latency

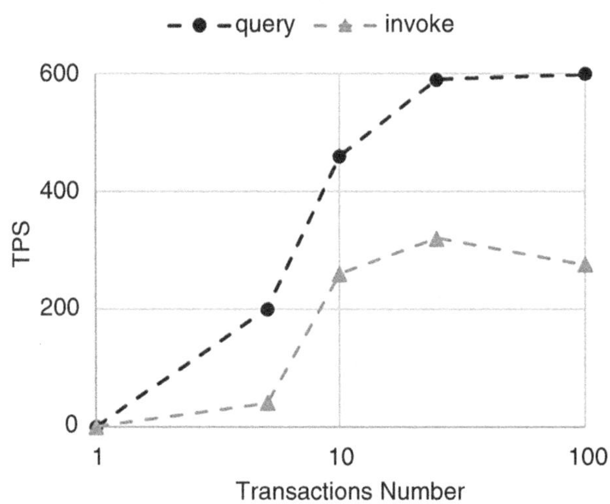

Fig. 8 Throughput

Considering the set of transactions from 10 to 100, the average latency is compared for each complete invoke function and at a different number of the involved peers, is evaluated and shown in Fig. 9. The average latency is essentially constant with the increase in the number of peers, but with the increase of transactions, its value considerably increases, increasing by an order of magnitude.

The scalability of this system can be tested, by adding new organizations to the system and evaluating the same previous parameters about execution time, latency, and throughput, and also varying the number of peers. In this case study, the network

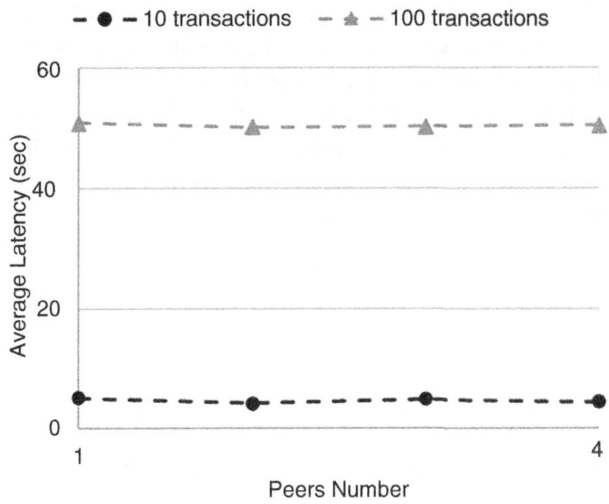

Fig. 9 Average latency at the boundary conditions about transactions, varying peers' number

works fine until the addition of other eight organizations which have two peers for each. After these conditions, the latency of the system reaches values too high to be used in a traceability system, i.e., exceeding the hour when the total number of organizations is 13.

Up to this limit value of the number of the organizations, a new organization can be added to the network, without a reset, and the network workload results to be unsustainable both in terms of control plane and data plane workloads.

Future analysis of them could be interesting to measure more precisely the workload impact for a single specific agri-food supply chain.

4 Conclusion

The agri-food supply chain was defined as a complex SOS and the counterfeiting of its products is really a problem for the consumers' health. Therefore, a multilayer solution for traceability was designed, with the combination and integration of different technologies, which are IoT sensors and blockchain technology. The possibility to use IoT devices in the edge and cloud environment provides the advantages of an automatic generation of the devices' list into the system and flexibility in the IT infrastructure, without further costs besides of cloud. IoT devices are installed into the critical points identified along the entire supply chain and some of these data – collected in the cloud environment – are registered in the blockchain technology through a specific smart contract or chaincode written in Go Language, called "Foodchain." The blockchain platform used in this work was an open-source platform – called Hyperledger Fabric – installed on the Google

Cloud environment. The main advantage of this proposed solution is represented by its scalability. This last claim was proved through the test of performances on the blockchain and executed, thanks to the Hyperledger Caliper tool. The system seems to get useful solutions in terms of latency and throughput, if it has a maximum of 12 organizations inside it. With the addition of the 13th eventual organization in the blockchain network, the value of the network workload is unsustainable both in terms of the control plane and data plane.

References

1. D.D. Walden and others, Systems engineering handbook: a guide for system life cycle processes and activities, in *INCOSE 2015* (2015)
2. T. Bosona, G. Gebresenbet, Food traceability as an integral part of logistics management in food and agricultural supply chain. Food Control (2013). https://doi.org/10.1016/j.foodcont.2013.02.004
3. I. Šenk, G. Ostojić, L. Tarjan, S. Stankovski, M. Lazarević, L.M. Camarinha-Matos, S. Tomic, P. Graça, Food product traceability by using automated identification technologies (2013). https://doi.org/10.1007/978-3-642-37291-917
4. K. Demestichas, N. Peppes, T. Alexakis, E. Adamopoulou, Blockchain in agriculture traceability systems: a review. Appl. Sci. (2020). https://doi.org/10.3390/app10124113
5. M. Bertolini, M. Bevilacqua, R. Massini, FMECA approach to product traceability in the food industry (2004). https://doi.org/10.1016/j.foodcont.2004.09.013
6. Z. Dong, C. Ma, Y. Wang, Z. Liu, Food information traceability system based on fabric, in *Proceedings of the 2020 International Conference on Aviation Safety and Information Technology* (2020). https://doi.org/10.1145/3434581.3434685
7. S. Islam, J.M. Cullen, Food traceability: a generic theoretical framework. Food Control (2021). https://doi.org/10.1016/j.foodcont.2020.107848
8. A. Tan, P. Ngan, A proposed framework model for dairy supply chain traceability. Sustain. Futur. (2020). https://doi.org/10.1016/j.sftr.2020.100034
9. A. Falcone, C. Felicetti, A. Garro, A. Rullo, D. Saccà, PUF-based smart tags for supply chain management, in *ACM International Conference Proceeding Series* (2021). https://doi.org/10.1145/3465481.3469195
10. J. Qian, L.F. Ruiz-Garcia, J.I. Beilei Robla Villalba, U. McCarthy, B. Zhang, Q. Yu, W. Wu, Food traceability system from governmental, corporate, and consumer perspectives in the European Union and China: a comparative review. Trends Food Sci. Technol. (2020). https://doi.org/10.1016/j.tifs.2020.03.025
11. European Parliament and Council Regulation (EC) No 178/2002. (2002). Available online: https://publications.europa.eu/en/publication-detail/-/publication/b18e2753-3f89-4e5e-8b03-6c6a9d340ac1/language-en
12. J.F. Galvez, J.C. Mejuto, J. Simal-Gandara, Future challenges on the use of blockchain for food traceability analysis. TrAC – Trends Anal. Chem. (2018). https://doi.org/10.1016/j.trac.2018.08.011
13. R. Kamath, Food traceability on blockchain: Walmart's Pork and Mango Pilots with IBM. J. Br. Blockchain Assoc. (2018). https://doi.org/10.31585/jbba-1-1-(10)2018
14. K. Biswas, V. Muthukkumarasamy, W.L. Tan, *Blockchain Based Wine Supply Chain Traceability System.* Future Technologies Conference (FTC). Vancouver, Canada. (2017)
15. M.T. Gaudio, S. Chakraborty, S. Curcio, Agri-food supply chain traceability: a multi-layered solution. *2022 IEEE International Conference Dependable, Autonomic Security Computer International Conference Pervasive Intelligence Computer. International Conference Cloud*

Big Data Computer. Intnational Conference Cyber Science Technological Congress (2022). https://doi.org/10.1109/DASC/PiCom/CBDCom/Cy55231.2022.9928017
16. M. Belotti, N. Božć, G. Pujolle, S. Secci, A Vademecum on blockchain technologies: when, which, and how. IEEE Commun. Surv. Tutorials (2019). https://doi.org/10.1109/COMST.2019.2928178
17. A Blockchain Platform for the Enterprise – Hyperledger-Fabricdocs Master Documentation Available online: https://hyperledger-fabric.readthedocs.io/en/release-2.3/. Accessed 24 Dec 2022
18. L. Foschini, A. Gavagna, G. Martuscelli, R. Montanari, Hyperledger fabric blockchain: chaincode performance analysis, in *ICC 2020–2020 IEEE International Conference on Communications* (2020). https://doi.org/10.1109/ICC40277.2020.9149080
19. Hyperledger Caliper – Hyperledger Foundation Available online: https://www.hyperledger.org/use/caliper. Accessed 24 Dec 2022

The Role of Federated Learning in Processing Cancer Patients' Data

Mihailo Ilić, Mirjana Ivanović, Dušan Jakovetić, Vladimir Kurbalija, Marko Otlokan, Miloš Savić, and Nataša Vujnović-Sedlar

1 Introduction

In contemporary society with constantly growing and turbulent economies, a range of serious epidemic/pandemic situations, and an increasing number of various social conflicts, people are under very stressful living conditions. Dynamic life and numerous everyday activities are significant factors in triggering critical health problems and serious diseases like cancer, cardiovascular and neurological diseases, and others. On the other hand, the latest cancer statistics[1] show that the absolute number of people living with cancer is constantly increasing. Cancer is a specific disease and there are a number of factors that affect its treatment and quality of life of cancer patients. Accordingly, it is necessary that big and important health and medical stakeholders and multidisciplinary teams try to find ways to develop supportive services to make life of such patients easier and bearable.

Information and communications technology (ICT) is a highly developing area, especially with AI/ML; use of wearable devices, sensors, and IoT; efficient techniques for processing big data; and others, which are very promising to cancer patients as they support the development of reliable and sophisticated solutions/services for before, during, and after treatment.

[1] doi.org/10.1016/S1470-2045(16)00091-7 – Article on cancer statistics.

M. Ilić (✉) · M. Ivanović · D. Jakovetić · V. Kurbalija · M. Otlokan · M. Savić
N. Vujnović-Sedlar
Faculty of Sciences, University of Novi Sad, Novi Sad, Serbia
e-mail: milic@dmi.uns.ac.rs; mira@dmi.uns.ac.rs; dusan.jakovetic@dmi.uns.ac.rs; kurba@dmi.uns.ac.rs; otlokanmarko@pmf.uns.ac.rs; svc@dmi.uns.ac.rs; natasa.vujnovic@dmi.uns.ac.rs

C. Savaglio et al. (eds.), *Device-Edge-Cloud Continuum*, Internet of Things,
https://doi.org/10.1007/978-3-031-42194-5_4

For successful employment of powerful AI/ML/FL approaches, it is necessary to have big and rich medical datasets and to find ways to collect the wide spectra of patients' complex data. Apart from standard clinical data, different health parameters should also be collected from wearable devices, sensors, everyday activities, nutritional data, environmental data, and so on. Patients' feedback obtained through different validated questionnaires that depict the particular QoL parameters is also very important. All kinds of patients' data are necessary for training and obtaining reliable predictive AI/ML/FL models [1]. Achieved results after successful processing of patients' data should be presented to doctors in an understandable way [2] so that they can be able to produce better explanations and recommendations to improve patients' health status.

The EU-funded H2020 project ASCAPE[2] is focused on the use of contemporary techniques in big data, artificial intelligence, and machine learning as a support to cancer (breast and prostate) patients in keeping and even improving their quality of life parameters [3, 4], by leveraging both cloud and edge intelligence techniques [5, 6]. As an outcome of the ASCAPE consortium[3] efforts, a powerful AI/ML/FL open infrastructure is implemented which will offer great opportunities for clinicians to make adequate personalized treatments and suggest interventions for patients [7]. The University of Novi Sad (UNSPMF), as a member of the consortium, has a crucial role within the ASCAPE project in implementing ML and FL aspects of ASCAPE architecture. The main intention of the chapter is to give a brief overview of the ASCAPE architecture but also to present key aspects of federated learning as an original scientific contribution of the UNSPMF team.

The rest of the chapter is organized as follows. In Sect. 2 several related projects are briefly presented and compared to results of our project. Section 3 considers key aspects of the federated learning paradigm. In Sect. 4 general overview and functionalities of ASCAPE architecture are presented. The central part of the chapter is Sect. 5 which brings theoretical and implementation aspects and describes the way in which FL is incorporated within ASCAPE architecture. The last section contains concluding remarks.

2 Related Projects and Work

In recent years, the European Commission [8] has been financing a number of large projects with focus on health, medical, and well-being domains. A significant number of them are devoted to the development of intelligent architectures and adequate services for different types of cancer in order to improve patients' QoL issues. Several projects related to ASCAPE will be briefly presented and compared to our approach.

[2] ascape-project.eu – Website of the ASCAPE project.

[3] ascape-project.eu/the_project/consortium – The ASCAPE consortium.

A noteworthy project is BD2Decide [9] which is devoted to the development of a more precise prognostic prediction in cancers of the head and neck region (HNC). Within the project, a decision support system (DSS) has been developed that combines existing data and advances of personalized prognostic models with technologies for data visualization to decide about HNC patients' treatment. Several data management components and functionalities are developed similar to the ASCAPE project. However, in comparison to ASCAPE, this system relies on a centralized infrastructure and collecting data from the various hospitals in a central repository. In our approach, a high priority is given to security, privacy, and the protection of patients' data in a distributed environment. An innovative aspect of our approach is based on the use of advanced techniques like differential privacy (DP) and homomorphic encryption (HE) within federated learning (FL) settings.

ONCORELIEF [10] is another project focusing on patients with two kinds of cancer, acute myeloid leukemia and colorectal cancer. The main achievement of the developed system is the improvement of posttreatment health status, increase of overall well-being, and follow-up care of cancer patients. The proposed framework includes rather standard components for data processing, using AI models and providing personalized support for patients through an intuitive smart digital assistant (Guardian Angel). However, the developed framework missed to employ the distributed style of processing datasets from different health institutions and use of contemporary federated learning. Contrary, the ASCAPE architecture is highly based on federated deep learning and distributed data processing applying nowadays popular and reliable techniques for security and preservation of patients' data.

The FAITH project [11] – a federated artificial intelligence solution for monitoring mental health status after cancer treatment – has some similarities with ASCAPE. The project result is an AI application based on FL that remotely identifies depression markers in people who have undergone cancer treatment (lung and breast cancer). Within the system data on lifestyle, clinical and behavioral data are also considered. Similarly, as in ASCAPE, this approach is based on FL and training global AI models that will be used in hospitals on their local data. However, in our approach, 15 QoL aspects for breast cancer and 12 QoL aspects for prostate cancer are considered not only depression as in FAITH project.

3 Different Aspects of Federated Learning

Federated learning is a recently proposed learning paradigm wherein a number of users (clients, devices, data owners) collaborate with the help of a central coordinator (server) in order to produce a joint ML model of common interest; see, e.g., [12, 13]. The joint model harnesses data from all users; however, the data is not shared during the learning process, but only trained models, or model increments. In more detail, a FL system consists of multiple users and a FL server. Users typically do not communicate one with another, but only with the server. The ML training process proceeds iteratively, over a number of epochs. At each epoch, the server transmits the current ML model (e.g., a neural network) to all (or a subset of) users. Subsequently, the users perform model updates with respect to their local data and transmit the updated local models, or model increments, to the server. Next, the server performs an aggregation step (averaging of the received local models) to fuse the model contributions from different users. The process then continues iteratively.

Such feature of FL makes it attractive from multiple perspectives. First, as the data is kept locally at each user, data privacy is improved over conventional learning systems that assume that raw data is transferred to a single location for learning. Second, when the data volumes are much higher than model sizes, FL also improves system efficiency from the perspective of communication traffic.

Depending on the background application and the type of the system of interest, FL approaches are usually categorized into two types: cross-device FL and cross-silo FL [13]. With cross-device FL, there is typically a huge number of users, and the users (devices) are usually low power, with limited computational and storage capabilities. With cross-silo FL, the users typically correspond to more powerful entities (servers) that may belong to different stakeholders. While the design of FL systems incurs a number of challenges as discussed below, some of those challenges are more relevant to a specific FL system type (cross-silo or cross-device). For example, data privacy may be more relevant with cross-silo FL, while communication efficiency may be more important with cross-device FL systems. We next discuss the FL literature with the relevant research trends and directions, considering several key challenges (dimensions): algorithmic aspects, model personalization, and communication efficiency.

Algorithmic aspects – From the mathematical point of view, model training via FL corresponds to execution of an iterative distributed optimization algorithm that minimizes a loss function, e.g., the empirical loss function based on the data arising from all users. The choice and design of this FL solver play an important role in the overall performance of a FL system. A widely used solver is the federated averaging (FedAvg) [14]. With FedAvg, at each epoch, the server selects K out of N users uniformly at random; the selected users perform a fixed number, say M, iterations of the stochastic gradient descent (SGD) with respect to their local data. Subsequently, the active users send their updated models to the server; the server then averages the received models. FedAvg shows good performance over a number of FL tasks. However, it suffers from bias toward users' local data when the number of local iterations M is taken to be too large. To this end, a method termed FedProx has been proposed that introduces proximal terms at local users' updates that better control the differences between local users' updates and the global model [15]. In addition in [16] is proposed a method that can better control the bias incurred when users exhibit heterogeneous capabilities and hence perform different numbers of local updates. In [17] is proposed FedHybrid, a method that enables different types of updates (e.g., gradient descent vs. Newton step) at different users, depending on their computational capabilities, in order to better harness the system heterogeneity. The fusion of weights from different ML models is proposed in [18] through a process called FDEML, which yields a new generalized model with better accuracy.

Model personalization – In a conventional FL setting, all users wish to obtain a common ML model trained on the data from all users. However, even though related to each other, the model requirements of different users can be slightly different, so that the trained model exhibits high test accuracy on local data of a given user, after training. For this reason, various personalized FL approaches have been proposed, which simultaneously enable model generalization, i.e., learning from all users' data, but also model personalization, so that the finally trained model for a given user is better adapted to its local data. A natural approach to personalized FL is twofold: first, a global model is learned via a standard FL algorithm, e.g., FedAvg. Then, each user makes a few training iterations (e.g., by performing SGD) with respect to its own local data. In more advanced variants, several approaches to personalized FL have been proposed, including multi-task learning [19, 20], fine-tuning [21, 22], and knowledge distillation [23–25].

Communication efficiency – With FL training, at each training epoch, at the uplink side, active users need to transmit their local models or local model increments to the server. Similarly, at the downlink side, the server transmits the global model to all (or to the active subset of) users. This process may incur a significant communication cost and also lead to a communication bottleneck at the server, especially when model sizes are huge. Indeed, huge models need to be trained with many modern deep learning applications. Therefore, it is desirable to reduce the communication overhead incurred while training a FL model. To this end, several strategies have been proposed. References [26–28] consider approaches where the messages communicated are quantized to a lower precision and projected to a lower dimension and hence incur lower communication costs. On the other hand, the authors of [29–31] consider optimized user sampling. That is, with these approaches, the server attempts to select the most informative users at each training epoch. Finally, references [32–36] consider one-shot methods that explore the best possible training schemes assuming that only one round of communication between the server and the users can be performed. Other approaches to communication efficiency include gradient sparsification [37, 38] and local methods [39–41].

4 ASCAPE Architecture

The ASCAPE architecture incorporates cloud/edge approach in order to make innovative AI/ML/FL models available to doctors and patients while preserving the privacy of patients' sensitive data. The first prototype of ASCAPE architecture was developed by a combined effort of all consortium members of the ASCAPE project [3, 4, 7, 42].

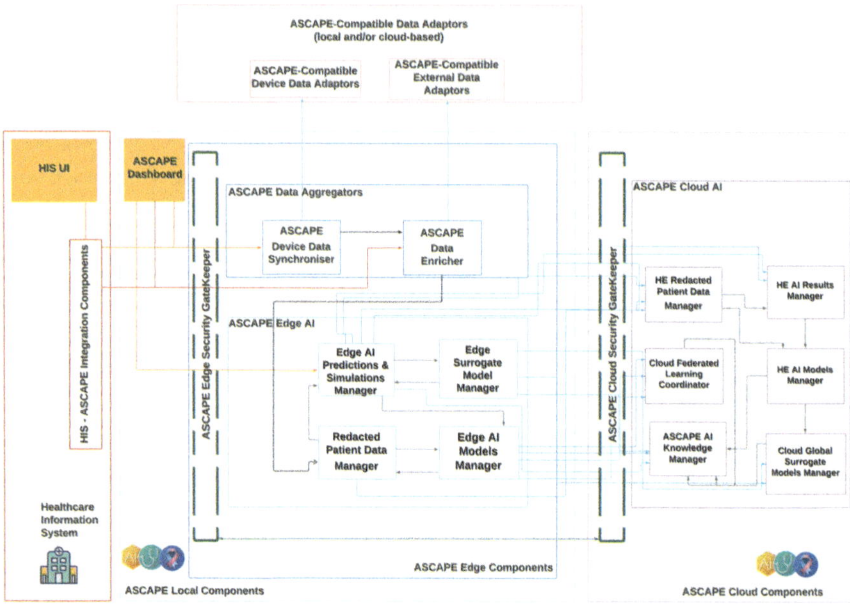

Fig. 1 The ASCAPE architecture

4.1 General Overview

The system is based on microservices as this paradigm fits naturally with processes within the ASCAPE framework [3]. All software components of the ecosystem form three distinct modules: the ASCAPE edge and cloud, external data aggregators, and pilot site-specific integration components [3, 7] (see Fig. 1).

ASCAPE edge and cloud – The edge components are located at the hospital and their main task is to synchronize the data collected from the healthcare information systems (HIS) and external data aggregators. These components also enable the training of local and global models while preserving data privacy. Finally, they offer prediction making functionality.

External data aggregators – The main task of these components is to collect data, which is not collected directly by the HIS and includes weather data, data collected by wearable devices (Fitbit), etc.

Pilot site-specific integration components – The main task of these components is the collection and transfer of data to the ASCAPE edge components for use in training and inference actions.

These modules support the following workflow: data ingestion, predictive model training, personalized AI predictions and simulations, and user interaction.

Data ingestion – The process of transferring data from the HIS to the ASCAPE edge component installed at the healthcare institution. Conversion to a suitable format for use in the ASCAPE environment must be executed first. Moreover, data enrichment may be carried out by including data not collected by the HIS, like weather data.

Predictive model training – The aggregated data may be used to train numerous ML/FL models. Three distinct cases are supported: local model training, global federated model training, and training of homomorphically encrypted global models. Local model training covers a wide variety of algorithms like support vector machines, random forest classifiers/regressors, etc. All models are validated against local validation data. In the case of local model training, only the model which performed the best on validation data is persisted and used later on for inference requests.

Personalized AI predictions and simulations – Personalized QoL predictions can be made for individual patients upon requests made from doctors. Inference requests are made to all models: local, global federated, and homomorphically encrypted. The effects of various interventions can also be simulated by using the available models.

User interaction – Doctors are granted access to the system through the ASCAPE dashboard. They are able to view and update patients' info, run predictions based on this data, and suggest further treatment.

Apart from the main goal of providing support to doctors using advanced ML methods, data security and preservation of privacy are crucial. Edge components are deployed at the medical institutions/hospitals (edge nodes), where integration with the existing HIS is achieved through ASCAPE integration components. Through this integration, patient data is aggregated and enriched with external ASCAPE compatible data. It is done locally at the healthcare institution due to data security.

Following this, local ML models can be trained which then provide predictions for each patient upon requests made by doctors. Patient data along with QoL predictions and simulations are made available to doctors through the ASCAPE dashboard in a user-friendly manner based on explainable AI (XAI) techniques. Edge computing allows healthcare providers to protect the privacy of patients' data while also assisting doctors with proposing treatments for their patients.

The ASCAPE cloud also enables all members of the ASCAPE ecosystem to exchange knowledge. Institutions can collaborate and train global models by using two privacy preserving ML methods.

Federated Deep Learning – By participating in FL, institutions train deep models on local data, after which they share the learned weights with the ASCAPE cloud. The FL Manager aggregates all of the received weights, after which it sends them back to the edge nodes for further training. Intermediate global models are persisted at the cloud and are made available to all members of the ecosystem.

Homomorphic Deep Learning – Patient data is homomorphically encrypted and uploaded to the cloud. Global models trained on this kind of data produce homomorphically encrypted outputs, which can be decrypted by using the secret key known only to the edge node.

4.2 Data Management and Role of ML in ASCAPE

The main idea within ASCAPE framework is to apply knowledge derived from existing cancer patients' datasets, who are potentially treated in different clinics and countries, and apply it to new cancer patients. This knowledge transfer will be realized through various ML/FL techniques. ML techniques analyze and process existing patients' data in order to learn and extract useful knowledge and patterns. However, before application of ML techniques, the data collected from the HIS should be prepared (ML data preprocessing) to better match ML algorithm.

To test and tune ML algorithms and necessary ML data preprocessing tasks, we used two existing datasets from the Örebro clinic partner from Sweden which contains datasets with anonymized medical data of patients diagnosed with breast cancer and with prostate cancer [42]. More precisely, two retrospective datasets were suitable for these tasks: breast cancer (BcBase) dataset (18988 rows $\times$ 47 columns) and prostate cancer Örebro (ORB) dataset (2466 rows $\times$ 124 columns). The BcBase measures only four aspects of QoL (anxiety, depression, insomnia, and pain) in the form of true/false values. This dataset is appropriate for the testing of classification algorithms. On the other hand, ORB dataset measured QoL with LISAT-11 Quality of Life Score which unify many QoL aspects into one numerical value. This dataset is used for testing of regression tasks. Moreover, these LISAT-11 values are measured in several time periods (36/60/120 months after diagnosis), so several datasets are extracted for each time value.

The first preprocessing task in this ML pipeline is missing value imputation (MVI). All data collected from clinics contained some amount of missing data, so this task is necessary since ML algorithms cannot work precisely with missing information. For the MVI, we used three algorithms from a Python library scikit-learn: simple imputer, iterative imputer, and imputation based on nearest neighbor. Simple imputer fills missing values within a single attribute by using the mean of the attribute's existing values. Iterative imputer creates a regression model for each attribute and then fills missing values of an attribute with predictions obtained from the attribute's regression model. A regression model for a single attribute is fit on instances with non-empty value for that attribute. In nearest neighbors' imputation, a Euclidean distance metric is used to find the nearest neighbors. Each missing feature is imputed using values from k (parameter of the method) nearest neighbors that have a value for the feature. The features of the neighbors could be averaged uniformly or weighted by distance to each neighbor.

We conducted a systematic evaluation [43] of all three imputation methods (also for different values of parameter k in nearest neighbors' imputation). In some cases, the simple imputer is better, while in some other cases, iterative imputer has better performance. The quality of the kNN imputer depends on its parameter k. Since the differences are quite small, and since the simple imputer is significantly faster than the other two, our final decision was to use the simple imputer.

The next preprocessing step is outlier detection (OD) and removal. Outliers are dataset instances that differ too much from the majority of other instances. This task is not obligatory, but outliers can negatively influence performance of AI models and should sometimes be removed prior to AI model training. For OD, we performed the following methodology: an autoencoder is trained on the whole dataset, and then the instances with the reconstruction losses higher than some user-defined threshold on trained encoder are removed from the dataset.

Extensive experiments were performed [43] on retrospective datasets in order to tune the value of threshold for OD. Almost all versions with OD performed better than without OD in the sense of ML accuracy. Finally, the value 1 of the user-defined threshold parameter gives the best performance for all regressors. However, this threshold value means the rejection of 17.66% of the data points in the training data which is a significant number. Therefore, in the initial phases of utilization of ASCAPE framework when there will be less amount of data points available, we will recommend a higher value of this parameter to provide more data points for model training. In the later phases, when a significant number of data points will be available, this value (1) of threshold will be applied.

The last preprocessing task in the ASCAPE ML pipeline is differential privacy (DP) which represents the main privacy preserving technique. The privacy of the data is one of the main objectives in ASCAPE since we are working with sensitive patient data. DP protects patients' data from so-called model-inversion attacks where some part of patients' data can be revealed from the results of ML models. DP is based on the idea that the outcome of the query posted to a protected database (or ML model) is essentially equally likely independent of whether any individual joins or refrains from joining the database. In such a way, the private data about a particular patient is protected since the system returns the result with the same probability whether a particular patient was involved in the analysis or not.

There are several mechanisms for applying DP, but in the ASCAPE project, we use the addition of Laplacian noise (a noise which follows Laplace distribution) to the training data. As expected, more noise means more privacy, but a lot of noise could lead to the reduction of model performance/accuracy. So, the main challenge in the introduction of privacy preserving techniques in ML models is the optimal balance between privacy and the utility of models. The amount of privacy is controlled through parameter ϵ. Smaller value of ϵ means stronger privacy, which is achieved by adding more noise to the training data. After experiments performed on retrospective data [43], we selected the value 0.7 for parameter ϵ, since it represents an optimal balance between the level of privacy and the accuracy of the trained ML models.

With completely prepared datasets, we performed deep ML analysis [44]. Six classification algorithms with BcBase datasets were performed: random forest (RF), support vector machine (SVM), naive Bayes (NB), K-nearest neighbor (KNN),

Table 1 F1 scores for classification on BcBase datasets

Dataset	RF	SVM	NB	KNN	DT	TFNN
Bc-Base Anxiety	0.48467	0.41099	**0.55204**	0.45848	0.51542	0.48052
Bc-Base Depression	0.47357	0.41252	**0.53362**	0.46204	0.5111	0.52346
Bc-Base Insomnia	0.52501	0.53348	**0.55364**	0.5024	0.51518	0.54068
Bc-Base Pain	0.48215	0.41665	0.51661	0.45699	**0.52117**	0.51366

Table 2 MAE values for regression in ORB datasets

Dataset	L. Reg.	R. Reg.	Lasso	ElNet	K. Ridge	SVM	RF	KNN	AdaBoost	TFNN
ORB-30-36	5.311	5.099	5.089	5.126	5.147	6.518	**5.033**	6.720	5.362	6.035
ORB-30-60	6.128	5.925	**5.886**	5.913	5.958	6.773	5.975	6.968	6.504	6.631
ORB-30-120	7.003	6.652	**6.478**	6.504	6.750	6.871	6.718	7.133	6.845	7.542
ORB-54-60	5.238	5.070	**4.840**	4.859	5.115	6.772	5.011	6.906	5.828	6.087
ORB-54-120	6.899	6.356	**6.180**	6.216	6.492	6.859	6.357	7.128	6.612	7.206
ORB-108-120	6.524	5.977	**5.437**	5.448	6.155	6.875	5.627	7.033	5.878	6.641

decision tree (DT), and neural network (TFNN). Additionally, ten regression algorithms were evaluated on ORB datasets: linear regression, ridge regression, kernel ridge regression, lasso regression, elastic net regression, SVM regression, RF regression based on 50 regression tree estimators, AdaBoost regression based on 50 regression tree estimators, KNN (k = 20) regression, and TensorFlow neural networks (ReLU, 50 epochs with early stopping, dropout, activity regularization).

Eight evaluation metrics were used to assess classification algorithms: accuracy, F1 score, precision of the positive class, recall of the positive class, precision of the negative class, recall of the negative class, macro-averaged precision, and macro-averaged recall. Also, four evaluation metrics were used for evaluation of regressors: mean absolute error (MAE), mean square error (MSE), coefficient of determination (R2), and Pearson's correlation coefficient (PC). However, due to the space limitations here, we show results only for F1 score (Table 1) and MAE (Table 2). The results for other evaluation metrics are in line with these results.

The results for classification are not particularly good since the F1 score for blind guessing would be at the level of 0.5. The NB classifier is the best for three datasets, while the SVM classifier is the worst again for three out of four datasets. A possible interpretation of these rather bad results is that variables (mainly related to the breast cancer diagnosis) do not contain the necessary information relevant for the prediction of the above target variables which measures patients' QoL.

The regression results are much better. Lasso regressor is the best, with winning results in five out of six datasets, while kNN regressor is the worst. However, MAE results in interval [4.84, 6.47] are quite better than classification results since the target variable (the LISAT QoL index) is in the range [11, 66]. An interesting finding here is that the short-time predictions (e.g., ORB-30-36 dataset predicts value of QoL in month 36 on the basis of data from the first 30 months) are quite better than long-time predictions (e.g., ORB-30-120).

5 Federated Learning Within ASCAPE Architecture

A key result of the ASCAPE project and the implemented architecture is the components which support federated learning. This privacy preserving method is crucial in allowing medical institutions to share knowledge and construct global ML models which are made available to all members of the ASCAPE system. In the rest of the section, we will present FL styles adopted in the ASCAPE architecture along with technical and implementation aspects.

5.1 Federated Learning Approaches

The term *federated learning* applies to decentralized learning where multiple clients collaborate in order to train a global machine learning model. The key difference from standard decentralized learning is that clients never share their data with the server, nor with other clients participating in the learning process. In most cases, there is a central server orchestrating the training process; however, there exist fully decentralized implementations where edge nodes communicate between themselves in order to collaboratively train models.

FL implementations may take on numerous forms, depending on the environment and particular problem they are applied to. For instance, the number of participating clients plays a significant part in how this approach is implemented. Another variable is how the global model is updated by the participating edge nodes. If all edge nodes update the model simultaneously, this is called concurrent federated learning. Contrary to this, incremental federated learning sees edge nodes taking turns in updating the global model, one after the other. Aggregating all of the updates to form the next iteration of the global model may also be done in numerous ways by using different formulas.

The ASCAPE architecture is meant for use in medical institutions, where the number of edge nodes is in the tens or hundreds, i.e., cross-silo FL. Each edge node E_i participating in FL has its own corresponding local training set D_i, where $i \epsilon [1 \ldots k]$. ASCAPE supports two main global model updating strategies, and also a novel, more flexible, approach to updating these models (see Fig. 2).

Incremental federated learning – The FL model is improved incrementally from the first to the last client. The model M is created by E_1 and trained on D_1 after which it is saved in the ASCAPE cloud. Following this, edge node E_2 retrieves model M and updates it using D_2. The entire process is finished once E_k submits its updates to the global FL model.

Concurrent federated learning – All edge nodes $E_1 \ldots E_k$ train model M simultaneously. Training of M is split into rounds, where in most cases, a round is equivalent to an epoch. At the end of each round, edge nodes submit their updates to the federated learning coordinator in the form of updated hidden layer weights,

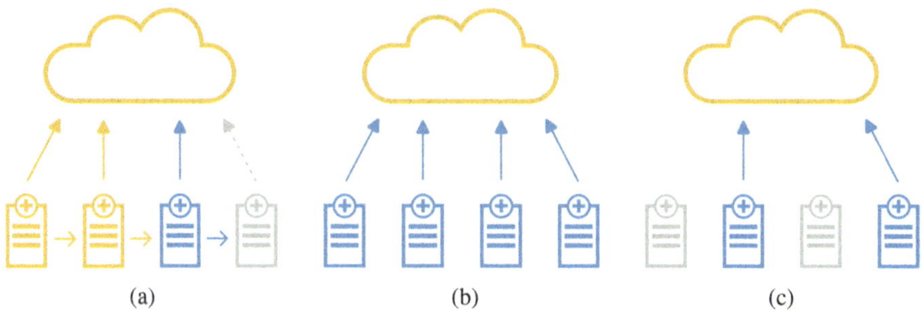

Fig. 2 Different FL model update strategies in ASCAPE. (**a**) Incremental FL. (**b**) Concurrent FL. (**c**) Semi-concurrent FL

where they are aggregated using the federated averaging [45] method. This updated model M is sent back to the edge nodes for a subsequent round of training. The process is repeated for an arbitrary number of rounds, until a viable model is produced.

Semi-concurrent federated learning – The ASCAPE framework also supports seamless switching between the previous two FL modes. Training can be started by a single healthcare institution, after which it can be joined by any number of other institutions during the training process.

Edge nodes contain their local data, the means of training both local and global federated models, along with components which generate predictions for patients' quality of life. The ASCAPE cloud enables training and storage of global federated models (see Fig. 3).

Regardless of the FL strategy used, all global FL models and their weights are stored in the ASCAPE cloud. These models are then made available for each and every member of the ASCAPE ecosystem to use. This ensures that even medical institutions which do not have an abundance of local data which they can use to train local models are still able to utilize knowledge gained elsewhere. Fueling this type of collaboration with a focus on preservation of privacy is one of the main goals of the ASCAPE project.

Global FL models are distinguished by their name which acts as a unique identifier. The first edge node that connects to the cloud in order to train a specific model is the one which defines its hidden layer structure. Before defining a new global FL model, an edge node may be configured to run architecture search on local data in order to determine the best hidden layer configuration to be used. The execution of this preparation step can be greatly beneficial during federated learning later on.

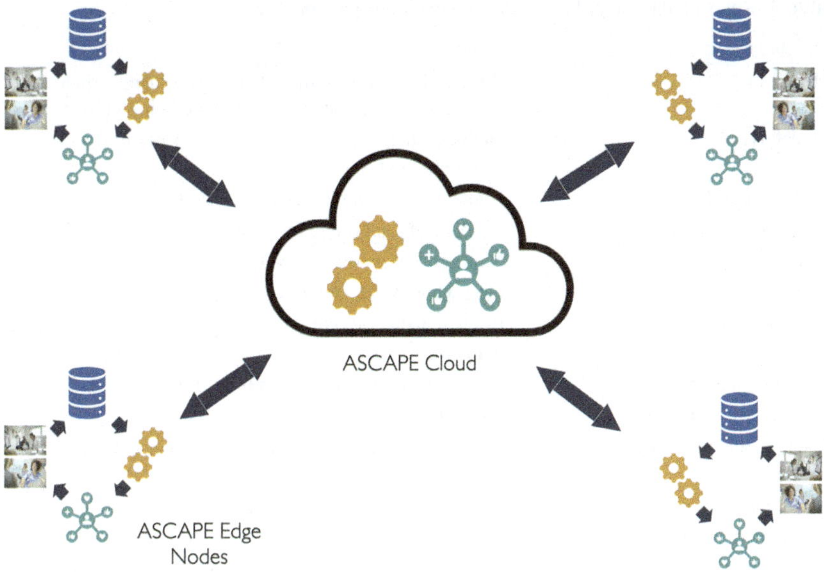

Fig. 3 Federated learning within the ASCAPE framework

5.2 *Technical Aspects of Federated Learning in ASCAPE*

The ASCAPE architecture provides healthcare institutions the possibility of training local models to be used by themselves and to share knowledge through the creation and training of global FL models. Confidential patient data never leaves its respective healthcare institution. As it was presented in previous sections, different classification and regression ML algorithms are incorporated in the architecture and several FL styles are adopted.

There are four main components of the ASCAPE architecture which support ML/FL workflows. The components located at the edge include the AI Models Manager and the Predictions and Simulations Manager, while the cloud components include the Federated Learning Coordinator and the AI Knowledge Manager. All of the components are implemented in Python using libraries and frameworks like Flask, Keras, scikit-learn, etc.

5.2.1 AI Models Manager

The AI Models Manager is the local edge component deployed at the HIS server. All local and global model training is executed at this component on local patients' datasets. The AI Models Manager is deployed as a Docker container to ensure that all instances located at different hospital servers have the same tools and libraries

Table 3 Relevant data stored in the AI Models Manager database

Variable name	Description
Model_id	Primary key for each model. This is a string which contains the name of the target value being predicted
Model_type	Defines whether the model is used for regression or classification tasks
Target_variable	Name of the variable being predicted
Total_predictor_variables	Number of total predictor variables
Predictor_variables	Names of all predictor variables
Created	Date when the model was created
Training_outcome	Boolean value marking if training finished successfully or with errors
Training_outcome_report	Additional comment related to training
Number_of_instances	Total number of training data points
Local_model_learning_algorithm	Local algorithm with the best performance during validation
Local_model_evaluation_score	The validation score of the best local algorithm
Local_model_evaluation_metric	The local algorithm validation metric used
Global_model_evaluation_score	The validation score of the global federated model
Global_model_evaluation_metric	The global model validation metric used

available during runtime. An API is implemented using the Flask framework which supports:

- Data import and training of both local and global federated models
- Serving local models for use in predictive tasks
- Serving validation results of local and global models

Local machine learning models are trained using the scikit-learn library, while deep federated machine learning models are trained using the Keras library. Validation of both types of models is also conducted after the training. Local models are validated through tenfold cross-validation, while for the global models, the dataset is split 80:20 for training and validation according to the Pareto principle.

All relevant data regarding model metadata and validation results is stored locally in an SQLite database (see Table 3).

5.2.2 Predictions and Simulations Manager

This edge component manages prediction tasks initiated by doctors for certain patients. QoL predictions are generated by both local models stored in the AI Models Manager and global federated models stored in the ASCAPE cloud. This component, like other edge components, is also deployed as a Docker container.

Table 4 Relevant ML model data stored in the Predictions and Simulations Manager database

Variable name	Description
Model_name	Name of the ML model
Path_to_model	Path to the model definition files stored locally
Input_dim	The format of the input data
Last_training_time	Last training time of the persisted model

The Prediction and Simulations Manager also exposes an API through Flask which is used by other components in the ASCAPE architecture and the ASCAPE dashboard. Doctors have the ability to:

- Upload patient data to get QoL predictions for the coming months
- Retrieve previous predictions for patients
- View statistics about patient's status presented a in user-friendly manner through the dashboard

This component also saves information locally. It persists historical prediction results in an SQLite database (see Table 4) along with the most recent global and local models to increase communication efficiency. Relevant data for AI models and prediction results can be seen in Tables 4 and 5.

5.2.3 Cloud Federated Learning Coordinator

The federated learning coordinator is a cloud-based component designed to facilitate the training of global federated machine learning models. Edge nodes communicate with the FL coordinator via web sockets since they enable efficient communication without interruptions. An open connection is needed in order to keep track of active edge nodes. Each model which can be trained in the ASCAPE ecosystem is uniquely identified by its name. A model training manager is instantiated for each of these models when edge nodes contact the federated learning coordinator in order to update a global model. The federated averaging algorithm is used to aggregate concurrent updates from multiple edge nodes. The impact of the updates from each edge node depends on the amount of data available at that node. Health institutions having more training data have a more significant impact on the global model than those with less data in concurrent federated learning.

$$\omega^{k+1} = \frac{\sum_{i=1}^{N} \omega_i^k * s_i}{\sum_{i=1}^{N} s_i} \tag{1}$$

The updated weights ω^{k+1} of the global federated model are calculated by averaging the updates ω^k of each of the N edge nodes at step k and scaling them by their impact s_i which is calculated by the number of patients they have compared to the rest of the edge nodes.

Table 5 Relevant prediction result data stored in the Predictions and Simulations Manager database

Variable name	Description
Target_name	Name of the target variable
Patient_id	Integer value of the patient id
Predicted_month	Character field for the month during treatment for which the prediction was made
Predicted_value	Float value of the local prediction
Predicted_value_global	Float value of the global prediction
Accuracy_local	Float value of accuracy of the local model
Accuracy_global	Float value of accuracy of the global model
AI_model	Foreign key of the local model

The training of global federated models consists of the following steps:

1. An edge node first contacts the ASCAPE cloud to check whether a model with the requested name exists.
   a. If it exists, the architecture and weights are sent back to the client which uses them to instantiate the current global model.
   b. If it does not exist, the edge node runs architecture search using local data and selects the best performing architecture which is sent back to the ASCAPE cloud to be stored.
2. The edge node executes 1 training step (usually corresponding to 1 epoch of training) using local data.
3. The updated model weights are sent back to the ASCAPE cloud to be aggregated with updates from other edge nodes active in the current training step.
4. Once aggregation is finished and the next iteration of the global model is ready, the ASCAPE cloud sends the newly updated global model back to all participating edge nodes.
5. The edge node repeats the process from step 2 onward until the performance of the global model is sufficiently reliable.

The federated learning coordinator persists the new global model after each epoch of training at the Cloud AI Knowledge Manager component.

5.2.4 Cloud AI Knowledge Manager

This cloud-based component manages the storage of global federated learning models in the ASCAPE architecture. During federated model training, after each step of the learning process, updates made by the edge nodes are aggregated by the federated learning coordinator and are sent to the AI Knowledge Manager in order to be persisted. These models are then made available to edge nodes to use in inference tasks.

Table 6 Relevant global federated ML model data stored in the AI Knowledge Manager database

Variable name	Description
Model_id	The name of the global model
Type	Determines whether the model performs regression or classification tasks
Last_training_time	Date and time of the last update made to the model by any ASCAPE edge node
Num_predictor_variables	The total number of input parameters
Predictor_variables	Names of all of the input parameters

Similar to the edge components mentioned in Sects. 5.2.1 and 5.2.2, this component exposes an API using the Flask framework in order to enable:

- Creation of new global federated models
- Updates of existing federated models
- Downloading of existing federated models for use in predictive tasks

This component stores global models along with their metadata in an SQLite database. Some of the relevant data stored in the database can be seen in Table 6.

6 Conclusion

Treatment and therapies prescribed by physicians for cancer patients usually negatively influence their everyday activities and QoL. Among several key priorities in the fields of medical/health informatics and engineering is the focus on the development of a range of intelligent systems and services that will help cancer patients to increase their health-related quality of life parameters. To achieve individual and personalized recommendations and interventions, it is necessary to have rich patient data records and reliable/powerful AI/ML predictive models trained and evaluated on adequate, big datasets.

Contemporary technologies and their combination offer great opportunities for clinicians to make adequate personalized treatments and suggest interventions that can increase patients' QoL. In this chapter, we briefly presented such an AI architecture which is based on contemporary AI/ML/FL technologies and at the same time provides security and privacy preservation of patients' data. In fact, we mostly focused on federated learning as one of the essential and specific aspects of ASCAPE architecture. More details on some other aspects of the ASCAPE architecture, such as data management, training and evaluation of ML models, and use of user-friendly dashboard based on several XAI techniques to present results of processed patients' data to physicians, can be found in [1, 3, 4, 7, 42, 44].

An additional benefit of the realized ASCAPE cloud/edge Architecture is that independent and other new health institutions can rather easily connect to the architecture and use developed AI/ML/FL functionalities and derived knowledge for their own purposes and cancer patients.

Acknowledgments This research was supported by the ASCAPE project. The ASCAPE project has received funding from the European Union's Horizon 2020 research and innovation program under grant agreement No. 875351.

References

1. M. Savić, V. Kurbalija, M. Ilić, M. Ivanović, D. Jakovetić, A. Valachis, S. Autexier, J. Rust, T. Kosmidis, Analysis of machine learning models predicting quality of life for cancer patients, in *Proceedings of the 13th International Conference on Management of Digital EcoSystems* (2021), pp. 35–42
2. A. Holzinger, G. Langs, H. Denk, K. Zatloukal, H. Müler, Causability and explainability of artificial intelligence in medicine. Wiley Interdiscip. Rev.: Data Min. Knowl. Disc. **9**(4), e1312 (2019)
3. K. Lampropoulos, T. Kosmidis, S. Autexier, M. Savić, M. Athanatos, M. Kokkonidis, T. Koutsouri, A. Vizitiu, A. Valachis, M.Q. Padron, ASCAPE: an open AI ecosystem to support the quality of life of cancer patients, in *2021 IEEE 9th International Conference on Healthcare Informatics (ICHI)* (IEEE, 2021), pp. 301–310
4. ASCAPE Deliverable – D1.1 Positioning ASCAPE's open AI infrastructure in the after cancer-care Iron Triangle of Health. https://ascapeproject.eu/node/57
5. S.U. Amin, M.S. Hossain, Edge intelligence and Internet of Things in healthcare: a survey. IEEE Access **9**, 45–59 (2020)
6. V. Barbuto, C. Savaglio, M. Chen, G. Fortino, Disclosing edge intelligence: a systematic meta-survey. Big Data Cogn. Comput. **7**(1), 44 (2023)
7. ASCAPE Deliverable - D4.1 Personalized interventions and user-centric visualizations. https://ascape-project.eu/node/120
8. European Comission Research and Innovation. https://commission.europa.eu/research-and-innovation_en
9. BD2Decide Project Website. http://www.bd2decide.eu
10. ONCORELIEF Project Website. https://oncorelief.eu
11. FAITH Project Website. https://www.h2020-faith.eu
12. T. Li, A.K. Sahu, A. Talwalkar, V. Smith, Federated learning: challenges, methods, and future directions. IEEE Signal Process. Mag. **37**(3), 50–60 (2020)
13. P. Kairouz et al., Advances and open problems in federated learning (2019). CoRR abs/1912.04977
14. H.B. McMahan, E. Moore, D. Ramage, S. Hampson, B.A. Arcas, Communication-efficient learning of deep networks from decentralized data, in *International Conference on Artificial Intelligence and Statistics* (2017)
15. L. Tian, A.K. Sahu, M. Zaheer, M. Sanjabi, A. Talwalkar, V. Smith, Federated optimization in heterogeneous networks. Proc. Mach. Learn. Syst. **2**, 429–450 (2020)
16. J. Wang et al., Tackling the objective inconsistency problem in heterogeneous federated optimization. Adv. Neural Inf. Process. Syst. Neurips **33**, 7611–7623 (2020)
17. X. Niu, E. Wei, FedHybrid: a hybrid federated optimization method for heterogeneous clients. IEEE Trans. Signal Process. **71**, 150–163 (2023)
18. S. Abbas, G.F. Issa, A. Fatima, T. Abbas, T.M. Ghazal, M. Ahmad, C.Y. Yeun, M.A. Khan, Fused weighted federated deep extreme machine learning based on intelligent lung cancer disease prediction model for healthcare 5.0. Int. J. Intell. Syst. **2023** (2023). https://doi.org/10.1155/2023/2599161

19. V. Smith, C.K. Chiang, M. Sanjabi, A.S. Talwalkar, Federated multi-task learning, in *Advances in Neural Information Processing Systems*, vol. 30 (Curran Associates, Inc., New York 2017)
20. F. Hanzely, P. Richtarik, Federated learning of a mixture of global and local models (2021)
21. A. Fallah, A. Mokhtari, A.E. Ozdaglar, Personalized federated learning with theoretical guarantees: a model-agnostic meta-learning approach, in *Advances in Neural Information Processing Systems 33: Annual Conference on Neural Information Processing Systems 2020, NeurIPS*, 6–12 Dec 2020 (2020)
22. K. Wang, R. Mathews, C. Kiddon, H. Eichner, F. Beaufays, D. Ramage, Federated evaluation of on-device personalization (2019)
23. G. Hinton, O. Vinyals, J. Dean, Distilling the knowledge in a neural network (2015). arXiv preprint arXiv:1503.02531
24. Y. Zhang, T. Xiang, T.M. Hospedales, H. Lu, Deep mutual learning, in *Proceedings of the IEEE Conference on Computer Vision and Pattern Recognition* (2018), pp. 4320–4328
25. I. Bistritz, A. Mann, N. Bambos, Distributed distillation for on-device learning, in *NeurIPS* (2020)
26. D. Alistarh, D. Grubic, J. Li, R. Tomioka, M. Vojnovic, QSGD: communication-efficient SGD via gradient quantization and encoding, in *Advances in Neural Information Processing Systems*, vol. 30 (Curran Associates, Inc., 2017). https://proceedings.neurips.cc/paper/2017/file/6c340f25839e6acdc73414517203f5f0-Paper.pdf
27. A. Reisizadeh, A. Mokhtari, H. Hassani, A. Jadbabaie, R. Pedarsani, FedPAQ: a communication-efficient federated learning method with periodic averaging and quantization, in *Proceedings of the Twenty Third International Conference on Artificial Intelligence and Statistics, ser. Proceedings of Machine Learning Research*, vol. 108, 26–28 Aug 2020, (PMLR, 2020), pp. 2021–2031. https://proceedings.mlr.press/v108/reisizadeh20a.html
28. A. Koloskova, S. Stich, M. Jaggi, Decentralized stochastic optimization and gossip algorithms with compressed communication, in *Proceedings of the 36th International Conference on Machine Learning, ser. Proceedings of Machine Learning Research*, vol. 97, 09–15 Jun 2019 (PMLR, 2019), pp. 3478–3487. https://proceedings.mlr.press/v97/koloskova19a.html
29. Y.J. Cho, J. Wang, G. Joshi, Client selection in federated learning: convergence analysis and power-of-choice selection strategies (2020). arXiv preprint arXiv:2010.01243
30. W. Chen, S. Horvath, P. Richtarik, Optimal client sampling for federated learning (2020). arXiv preprint arXiv:2010.13723
31. M. Ribero, H. Vikalo, Communication-efficient federated learning via optimal client sampling (2020). arXiv preprint arXiv:2007.15197
32. Y. Zhang, M.J. Wainwright, J.C. Duchi, Communication-efficient algorithms for statistical optimization, in *Advances in Neural Information Processing Systems*, vol. 25 (Curran Associates, Inc., 2012). https://proceedings.neurips.cc/paper/2012/file/e7f8a7fb0b77bcb3b283af5be021448f-Paper.pdf
33. N. Guha, A. Talwalkar, V. Smith, One-shot federated learning (2019). arXiv preprint arXiv:1902.11175
34. Y. Zhou, G. Pu, X. Ma, X. Li, D. Wu, Distilled one-shot federated learning (2020). arXiv preprint arXiv:2009.07999
35. S. Salehkaleybar, A. Sharifnassab, S.J. Golestani, One-shot federated learning: theoretical limits and algorithms to achieve them. J. Mach. Learn. Res. **22**(189), 1–47 (2021). http://jmlr.org/papers/v22/19-1048.html
36. D.K. Dennis, T. Li, V. Smith, Heterogeneity for the win: one-shot federated clustering, in *Proceedings of the 38th International Conference on Machine Learning, ser. Proceedings of Machine Learning Research*, vol. 139, 18–24 Jul 2021 (PMLR, 2021), pp. 2611–2620. https://proceedings.mlr.press/v139/dennis21a.html
37. S.U. Stich, J.B. Cordonnier, M. Jaggi, Sparsified SGD with memory, in *Advances in Neural Information Processing Systems*, vol. 31 (Curran Associates, Inc., 2018). https://proceedings.neurips.cc/paper/2018/file/b440509a0106086a67bc2ea9df0a1dab-Paper.pdf
38. J. Wangni, J. Wang, J. Liu, T. Zhang, Gradient sparsification for communication-efficient distributed optimization, in *Advances in Neural Information Processing Sys-*

tems, vol. 31 (Curran Associates, Inc., 2018). https://proceedings.neurips.cc/paper/2018/file/3328bdf9a4b9504b9398284244fe97c2-Paper.pdf
39. S.U. Stich, Local SGD converges fast and communicates little, in *ICLR 2019-International Conference on Learning Representations*, no. CONF (2019)
40. A. Khaled, K. Mishchenko, P. Richtarik, Tighter theory for local SGD on identical and heterogeneous data, in *Proceedings of the 41 Twenty Third International Conference on Artificial Intelligence and Statistics, ser. Proceedings of Machine Learning Research*, vol. 108, 26–28 Aug 2020 (PMLR, 2020), pp. 4519–4529. https://proceedings.mlr.press/v108/bayoumi20a.html
41. K. Mishchenko, G. Malinovsky, S. Stich, P. Richtarik, ProxSkip: Yes! Local gradient steps provably lead to communication acceleration! Finally!, in *Proceedings of the 39th International Conference on Machine Learning, ser. Proceedings of Machine Learning Research*, vol. 162, 17–23 Jul 2022, (PMLR, 2022), pp. 15,750–15,769. https://proceedings.mlr.press/v162/mishchenko22b.html
42. ASCAPE Deliverable – D2.4 ML-DL Training and Evaluation Report. https://ascape-project.eu/node/118
43. ASCAPE Deliverable – D2.3 ML/DL algorithms and Model Analysis and Report. https://www.ascape-project.eu/resources/deliverables
44. M. Savić, V. Kurbalija, M. Ilić, M. Ivanović, D. Jakovetić, A. Valachis, S. Autexier, J. Rust, T. Kosmidis, The Application of machine learning techniques in prediction of quality of life features for cancer patients. Comput. Sci. Inf. Syst. **20**(1), 381–404 (2023). https://doi.org/10.2298/CSIS220227061S
45. P. Kairouz, H.B. McMahan, B. Avent, A. Bellet, M. Bennis, A.N. Bhagoji, K. Bonawitz, Z. Charles, G. Cormode, R. Cummings et al., Advances and open problems in federated learning. Found. Trends®Mach. Learn. **14**(1–2), 1–210 (2021)

Scheduling Offloading Decisions for Heterogeneous Drones on Shared Edge Resources

Giorgos Polychronis and Spyros Lalis

1 Introduction

Drones can serve as cheap and easy-to-deploy aerial sensors, benefiting multiple applications. For instance, drones are already being used in smart cities [1], smart agriculture [2], and surveillance [3]. Besides simple data collection tasks where data processing and analysis takes place in the cloud after the mission is completed, there are cases where the drone has to process the data in situ and in a synchronous manner, before it moves to the next point of interest. However, small/cheap drones typically carry small and resource-constrained computing platforms. As a consequence, heavyweight computations can take a long time to complete and can significantly increase the total time it takes for the drone to complete the mission.

One possible way to reduce the mission time (or, equivalently, to increase the operational capability) of the drone is to offload such computations to more powerful nearby edge servers, instead of performing them onboard. However, this is not always feasible when the servers have limited capacity and there are several drones competing for these resources, and a purely opportunistic approach may result in an unfair usage of resources. To ensure that the available edge resources are fairly shared between the drones, it is important to plan the drones' paths in conjunction with an offloading schedule. The problem becomes even more challenging when the drones are heterogeneous, making it harder to achieve a balanced and fair schedule.

In previous work, we have proposed a greedy algorithm [4] and a meta-heuristic algorithm [5] to tackle this problem. In this book chapter, we study these algorithms for a wide range of mission scenarios with focus on the heterogeneity of the drones as well as in the heterogeneity of the missions to be completed by the drones. Our

G. Polychronis (✉) · S. Lalis
University of Thessaly, Volos, Greece
e-mail: gpolychronis@uth.gr; lalis@uth.gr

C. Savaglio et al. (eds.), *Device-Edge-Cloud Continuum*, Internet of Things,
https://doi.org/10.1007/978-3-031-42194-5_5

evaluation shows that the latter approach can evenly reduce the mission time of all drones regardless of the heterogeneity of both drones and missions and can improve the drones' mission times up to 21.8% versus a plan where all the computations are done locally on the drones.

The rest of the chapter is structured as follows. Section 2 discusses related work. Section 3 presents the system model. Section 4 describes the greedy algorithm, while Sect. 5 describes the meta-heuristic. These three sections reuse material from [4] and [5], which is briefly presented again here for completeness. Section 6 presents the evaluation of these algorithms for the case of heterogeneous drones. Finally, Sect. 7 concludes the chapter.

2 Related Work

In [6], a drone must stop at predefined waypoints and perform heavyweight computations. The drone reduces its mission time by opportunistically offloading its computations to an edge server when the offloading is beneficial. In [7], a drone-based system is used for tracking moving objects. The authors deal with the problem of deciding whether to offload a computation or execute it on the drone locally in order to minimize the response time. Unlike these works, which focus on runtime offloading decisions, we tackle the problem of planning the offloading decisions jointly with the drones' paths before the mission starts.

A large group of work considers drones as edge servers, trying to serve users by optimizing the trajectory of the drones and the associated resource allocation. Such a problem is studied in [8] where a UAV is used as a base station for providing video content to multiple users. The authors of [9] investigate a problem where mobile devices that are not covered by an infrastructure may offload their computation to an edge-mounted UAV. The goal is to minimize the overall energy needed for IoT devices to complete their tasks by making decisions about the offloading, resource allocation, and UAV trajectory. Both works split the problem in smaller subproblems that are solved iteratively until convergence is reached in the optimization criteria. In [10], the authors address the problem where a UAV is used as an edge server for multiple mobile devices in order to offload their tasks. The goal is to jointly optimize the UAV trajectory and the computation offloading decisions and resource allocation to minimize the average weighted energy consumption of the UAV and the mobile devices. As above, this work subdivides the problem into smaller parts that are solved in an iterative way. Our work follows a similar approach to reduce complexity, by separating the path optimization from the resource allocation aspect as each problem is hard on its own.

Another work that considers a UAV as a server is [11], where the UAV is used to serve content to moving vehicles in a highway, by moving to and hovering over different locations, sending and receiving data to/from clients, and using a data cache with a maximum capacity. The goal is to maximize the energy efficiency of the UAV by optimizing its movement, cache management, and data transfers.

In [12], multiple UAVs serve multiple IoT devices that need to offload their sensor data. The UAVs move to the proper locations in order to retrieve and process the data from the IoT devices. The goal is to serve all devices while minimizing the energy spent by the UAVs for flight and hovering. The main difference between all these problems and our work is that drones play the role of a server, whereas in our case, drones are the clients of more powerful edge servers.

The authors in [13] address the allocation of computing resources to services in a priority-aware way. They study two strategies: one that greedily considers services with higher priority and one that assigns bigger weights to high-priority services while preventing the starvation of services with lower priority. In [14], a hybrid fog and cloud-aware heuristic is proposed, which attempts to offload computationally heavy tasks to the cloud and communication-heavy tasks to the fog. The offloading priority is set based on the task deadlines (earliest deadline first), and the VM to which the tasks are offloaded is the one with the earliest estimated completion time. A similar approach is adopted in [15] to orchestrate dynamically arriving task workflows in the fog. A workflow consists of tasks which need to perform a certain amount of computation followed by an amount of communication to send the data produced to another task down the line. We investigate a different optimization objective than the above works, proposing corresponding heuristics.

The work in [16] shows that simple algorithms, which are based on congestion estimation, can be used for task allocation in fog nodes and outperform the current nearest-node algorithm, leading to improved node usage and smaller job delays. We also estimate the expected contention on servers to plan resource allocation, but, in our case, the offloading reservations and respective allocations are planned offline, before the execution of the missions. The authors in [17] addresses the problem of orchestrating bag-of-tasks jobs in a set of processors. The tasks in a bag can run parallel to each other, and a task may dynamically spawn another task if its QoS is not sufficient. The authors test three allocation techniques, RR (round-robin), 2RC (2 random choices), and SQ (shortest queue), for allocating processors to tasks. Our heuristics also allocate the available server resources to drones in a greedy way, but the decision is based on the expected completion time rather than the length of the server queues.

In [4], we have studied the joint path planning and offload scheduling problem, proposing a greedy algorithm with variations regarding the order in which the next planning decision is taken for each drone. In [5], we tackled an extension of this problem, for drones with energy constraints, which can switch batteries at special depot stations. To this end, we proposed a meta-heuristic algorithm that generates and compares between multiple randomly generated plans. Another version of the problem was studied in [18] for the case where there is uncertainty in the drone flight times, by introducing an online approach to dynamically update the schedule of each drone in a decentralized manner. However, the above works study the problem for homogeneous drones. In contrast, this work focuses on mission scenarios where the drones are heterogeneous in terms of flight capability as well as onboard processing capacity.

3 Model

This section captures the salient aspects of the problem in a more formal way. This makes it possible to describe the problem at hand as well as the desired optimization objective, in a concrete and unambiguous way.

3.1 Missions, Paths, and Flight Model

Let there be M drones, $d_m, 1 \leq m \leq M$. Each drone needs to complete its own mission, which includes visiting specific points of interest $p \in \mathcal{V}_m$ where the drone has to (i) perform some sensing and then (ii) process the collected data on the spot, e.g., to take further action if a problem is detected. Notably, we assume that data processing needs to be performed in situ, while the drone is positioned over the point of interest, before moving to the next one.

Let $senseT_m$ be the time that is needed for d_m to perform the required sensing. Afterward, the drone needs to process the sensor data. Let $comp_m$ be the type of computation that needs to be performed for this purpose. This determines both the amount of resources $res(comp_m, hw)$ and the amount of time $procT(comp_m, hw)$ required to perform the computation on a given hardware platform hw. Also, each computation $comp_m$ takes some data as input and produces some data as output, denoted as $comp_m^{in}$ and $comp_m^{out}$, respectively.

In order to visit its points of interest, each drone follows a path that is computed offline, before the drone starts its mission. Let P_m be the path computed for d_m, defined as a sequence of waypoints $P_m[i]$, $1 \leq i \leq \text{len}(P_m)$. The path must include all points of interest assigned to the drone:

$$\exists i : 1 \leq i \leq \text{len}(P_m) : P_m[i] = p, \forall p \in \mathcal{V}_m \tag{1}$$

Furthermore, we assume that every drone d_m starts its mission by taking-off from a depot station dep_m where it also needs to land at the end of the mission. We encode this in the drone's path by setting $P_m[1] = dep_m$ and $P_m[\text{len}(P_m)] = dep_m$. Without loss of generality, we assume that the depot is not among the points of interest, $dep_m \notin \mathcal{V}_m$.

Let $flyT_m^{i,i+1}$ be the time that is needed for the drone d_m to fly between two successive waypoints $P_m[i]$ and $P_m[i+1]$. This can be estimated as a function of the distance between these points, the default flight speed, the acceleration until the default speed is reached, and the deceleration as the drone approaches the target position where it needs to stand still. We assume that the drone flies between two points in a straight line, which is a realistic assumption for aerial vehicles.

Notably, weather conditions (wind or occasional gusts) in the area can affect the flight time between two points. In case such effects cannot be estimated, an approach like in [18] must be adopted. However, in this work, we assume that the flight times between two points can be estimated with sufficient accuracy.

3.2 Onboard Computation and Computation Offloading

Each drone d_m has an onboard computer with hardware platform hw_m, which can be used to perform $comp_m$ locally. In order for the drone to be truly autonomous, we assume that its onboard hardware platform has sufficient resources to run the computation in question, $res(hw_m) \geq res(comp_m, hw_m)$. Also, we assume that local data movement for $comp_m^{in}$ and $comp_m^{out}$ takes a negligible amount of time. Thus, the total computation delay when processing is performed on the drone is $compT_m = procT(comp_m, hw_m)$.

As another option, the computation can be offloaded from the drone to a nearby server in order to accelerate processing and reduce the mission time. We assume that each server is suitably prepared to provide such a computation as a service. For instance, the respective software can be packaged and shipped to the servers in the form of micro-VMs or containers, before the drones start their missions.

Let there be K edge servers, s_k, $1 \leq k \leq K$, statically placed at different locations in the wider area where the drones operate. Each server s_k may have a different hardware platform hw_k with computing resources $res(hw_k)$. Also, each server is accessible via a dedicated local wireless network with bandwidth bw_k and communication range $range_k$.

When the drone d_m visits a point of interest p which is in range of server s_k, it can offload its computation $comp_m$ to it. Then, the computation delay for the drone is

$$compT(m, k) = procT(comp_m, hw_k) + \frac{comp_m^{in} + comp_m^{out}}{bw_k} \tag{2}$$

taking into account the time needed to perform the computation on the server and transfer the respective input and output over the network. For convenience, let $compT(m, 0) = compT_m$ (the server identifier 0 means "no offloading").

Let $S_m[i]$ encode the offloading for d_m during its mission. More specifically, $S_m[i] = k$ if the drone shall offload $comp_m$ to s_k at $P_m[i]$, else $S_m[i] = 0$ if the drone shall perform the computation locally. Also, $S_m[i] = 0$ for $P_m[i] = dep_m$ as the drone does not perform any computation at the depot station.

Note that the drone may not be able to *immediately* offload its computation to $S_m[i]$ as soon as it completes sensing at $P_m[i]$. The reason is that server's resources may be used by other drones at that time. It is, however, possible for the drone to wait for the server to become available and then proceed with offloading. Let the extra waiting time for this be $W_m[i]$ for all points where $S_m[i] \neq 0$ ($W_m[i] = 0$ for $S_m[i] = 0$).

Based on the above, the time spent by the drone in order to perform the necessary sensing and computation at each point of interest $P_m[i] \in \mathcal{V}_m$ is

$$pT_m^i = senseT_m + W_m[i] + compT(m, S_m[i]) \tag{3}$$

and the time needed for d_m to complete the entire mission is

$$mT_m = \sum_{i=1}^{len(P_m)-1} flyT_m^{i,i+1} + pT_m^{i+1} \tag{4}$$

3.3 Server Resource Usage

Let matrix U encode the planned usage of the server infrastructure, where $U[k][m][t] = 1$ if s_k allocates resources to run the service for $comp_m$ at time t, else $U[k][m][t] = 0$. Note that the total amount of resources allocated on a server due to offloading cannot exceed its overall resource capacity:

$$res(hw_k) \geq \sum_{m=1}^{M} U[k][m][t] \times res(comp_m, hw_k), \forall k, t \tag{5}$$

We assume that the resources $res(comp_m, hw_k)$ associated with the service for performing $comp_m$ on s_k remain allocated only while the service is actually being used by a drone. More specifically, if d_m starts offloading its computation to s_k at time t_{start}, it will consume resources during the entire duration of the computation including the required input/output data transfers, $U[k][m][t] = 1, t_{start} \leq t \leq t_{start} + compT(m, k)$. Notably, we assume that an edge server cannot further offload a computation to another server.

3.4 Problem

Our goal is to exploit the available server infrastructure in order to reduce the mission time of all drones. More specifically, the problem we tackle is as follows.

Given drones $d_m, 1 \leq m \leq M$ with assigned points of interest $\mathcal{V}_m$ where they have to perform some sensing and data processing, and servers $s_k, 1 \leq k \leq K$ in the mission area that can be used for offloading the drones' computations, produce paths P_m and offloading plans S_m that minimize mT_m (as per Eq. 4) *in a fair way for all drones* while ensuring that the constraints in Eqs. 1 and 5 are satisfied.

We capture the fairness objective by using as a reference the makespan of a default path plan P_m^{def}, which is provided as input. The corresponding mission time

mT_m^{def} can be computed as per Eq. 4 for $S_m^{def}[i] = 0$, $1 \leq i \leq \text{len}(P_m^{def})$. Based on this reference, we set as the optimization target

$$\text{maximize } min_{m=1}^{M} \left(\frac{mT_m^{def} - mT_m}{mT_m^{def}} \right) \tag{6}$$

In other words, we wish to evenly maximize, across all drones, the relative reduction of their mission time, using the available server infrastructure for offloading.

4 Greedy Algorithm

The algorithm takes as input a default path P_m^{def} for each drone d_m, generated separately for each drone using a conventional TSP algorithm, which includes the required visitations at all points of interest $\mathcal{V}_m$, without any computation offloading. Note that we tackle applications where the drone does not have to follow a pre-specified path. It is merely required to visit all specified points of interest in any order. Thus, the P^{def} is merely indicative.

As a first step, the algorithm manipulates the default paths P^{def} so as to reduce the overall contention of the drones for the edge resources. This is done via function REVERSEPATHS(), which works in an iterative way, by randomly picking a drone and exploring whether a complete path reversal (visiting the points of interest in reverse order) would reduce the contention of the system. If so, the reversed path is adopted instead of the default (note that such a path reversal does not change the default mission time). The expected contention is roughly estimated based on the time periods where more than one drone need to perform computations at points of interest in range of the same servers, without considering any actual offloading.

The algorithm then incrementally introduces offloading in an iterative way. In each iteration, a drone is picked and the next visit is planned for it, updating the drone's state and server resource usage accordingly. If the end of the path is reached for a drone, it is removed from the set of candidates considered for the next iteration. Finally, the algorithm terminates when there are no candidates left. A structured description is given in Algorithm 1, which is briefly discussed below.

Function PICKDRONE() selects the drone to be considered in the current iteration of the algorithm. We experiment with three different policies, discussed in the following subsections. Function PLANNXTVISIT() plans the visit to the next point in the drone's path and updates its mission time. The possibility of offloading the drone's computation at that point is examined in the RESERVEOFFLOAD() function, which greedily finds via *nxtOffload()* the best available server and associated waiting time, which minimizes Eq. 3. If found, the resources needed for the computation are reserved on the server, and the drone's offload plan is updated accordingly.

We experiment with three policies, implemented as different versions of function PICKDRONE(): (i) round-robin (RR), (ii) current benefit (CB), and (iii) expected

Algorithm 1 Greedy algorithm

function MAIN($\mathcal{D}$, P^{def})
  $P \leftarrow$ REVERSEPATHS(P^{def})
  for each d_m **do**
    $curr_m, t_m \leftarrow 1, 0$ ▷ start from 1st waypoint, at time zero
    $S_m[i] = 0, 1 \leq i \leq \text{len}(P_m)$ ▷ no offloading, yet
    $W_m[i] = 0, 1 \leq i \leq \text{len}(P_m)$ ▷ no waiting times, yet
    $U[k][m][t] \leftarrow 0, \forall k, t$ ▷ no reservations, yet
  end for
  while $\mathcal{D} \neq \emptyset$ **do**
    $d_m \leftarrow$ PICKDRONE($\mathcal{D}$)
    PLANNXTVISTIT(d_m, U)
    if $curr_m = \text{len}(P_m)$ **then** ▷ end of mission
      $\mathcal{D} \leftarrow \mathcal{D} - d_m$
    end if
  end while
end function

function REVERSEPATHS(M, P^{def})
  $P \leftarrow P^{def}$
  for M **do**
    $m \leftarrow \text{pickRandom}(M)$
    $P' \leftarrow P - P_m + \text{reversePath}(P_m)$
    if $\text{calcContention}(P') < \text{calcContention}(P)$ **then**
      $P \leftarrow P'$
    end if
  end for
  return P
end function

function PLANNXTVISIT(d_m, U)
  $i \leftarrow curr_m$
  $t \leftarrow t_m + flyT_m^{i,i+1} + senseT_m$
  $k \leftarrow$ RESERVEOFFLOAD($d_m, t, i+1, U$)
  $t_m \leftarrow t_m + flyT_m^{i,i+1} + pT_m^{i+1}$
  $curr_m \leftarrow i + 1$
end function

function RESERVEOFFLOAD(d_m, t, i, U)
  $(k, t_{start}) \leftarrow \text{nxtOffload}(P_m[i], t, comp_m, U)$
  if $k \neq 0 \wedge t_{start} - t + compT(m,k) < compT(m,0)$ **then**
    $U[k][m][t] \leftarrow 1, t_{start} \leq t \leq t_{start} + compT(m,k)$
    $S_m[i] \leftarrow k$
    $W_m[i] \leftarrow t_{start} - t$
  end if
end function

benefit (EB). The RR policy simply picks the next drone in the candidate list (where drones are initially placed in random order) in a blind round-robin fashion. The CB policy picks the drone that has currently benefited the least from the available server infrastructure in terms of the relative reduction of its default mission time, thanks to offloading. More specifically, CB picks the d_m with the minimum $\frac{mT_m^{def} - mT_m^{upd}}{mT_m^{def}}$, where mT_m^{upd} is initially set equal to mT_m^{def} and is updated after a computation offloading is planned in the RESERVEOFFLOAD() function.

Finally, the intuition behind the EB policy is the following. Assume there are two drones, d_1 and d_2, with $len(P_1^{def}) < len(P_2^{def})$. Also, assume that a single server is accessible from all points of interest and that if one of the two drones offloads its computation to the server, then the other drone loses this offloading opportunity. Consequently, d_1 competes with d_2 for server usage during its entire mission, whereas d_2's mission includes a contention-free part where the drone can freely offload its computations to the server. Note that in this case, RR will evenly allocate the server to d_1 and d_2 during the period of contention, while CB will even tend to favor d_2 over d_1. Obviously, both approaches are not optimal with respect to the objective expressed in Eq. 6. A better approach would be to give (some) preference to d_1 during the period of contention since d_2 will be able to freely use the server for the remainder of its mission without any contention. To this end, one needs to consider the offloading opportunities and level of contention each drone will have during its mission.

More specifically, in each iteration, EB picks the drone d_m with the minimum $\frac{mT_m^{def} - mT_m^{exp}}{mT_m^{def}}$, where mT_m^{exp} is the expected mission time. This is estimated based on the computation offloading that has been planned so far, and the expected offloading decisions that will be taken in the future as part of the planning algorithm, for each point in the drone's path.

To determine the expected offloading decision at point $P_m[i]$ where d_m is expected to arrive at time t, first we estimate the potential contention $cont_{m,i}$ at this point, as the number of drones that are expected to arrive at (perhaps other) points of interest covered by the same servers at a time that is "close" to t (where offloading is still beneficial despite the potential waiting time due to server overload), divided by the number of servers that cover these points. Then, the probability for d_m to be able to offload its computation at $P_m[i]$ is

$$P_{m,i}^{offld} = \begin{cases} 1 & \text{if } cont_{m,i} \leq \Upsilon_i \\ \frac{w_{m,i} \times \Upsilon_i}{cont_{m,i}} & \text{otherwise} \end{cases} \tag{7}$$

with $\Upsilon_i = \frac{avgFlyT}{avgCompT_i} \times avgCapacity_i$, where $avgFlyT$ is the average flight time between waypoints (over all drones), $avgCompT_i$ is the average time needed by the servers reachable from $P_m[i]$ to perform the computations of the drones that can use these servers for offloading at the same time as d_m, and $avgCapacity_i$ is the average capacity of these servers. The factor $\frac{avgFlyT}{avgCompT_i}$ stands for the number of computations that can be performed by the servers sequentially before any drones that are in transit might require the server for offloading (thereby leading to a potential contention situation). Also, $w_{m,i} = \frac{len(P_m)-i}{len(P_m)}$ is a weight that is higher for the points of interest early in the drone's path and decreases for the ones further down the line. The rationale is for the offload estimation to be more pessimistic for the far future (if there is contention) as there is increased uncertainty about the decisions that will be taken by the algorithm at distant points.

Finally, the expected total mission time is estimated by using, at each point $P_m[i]$, Eq. 7 to compute the probability $P_{m,i}^{offld}$ for d_m to offload the computation to a server, or to perform the computation locally with probability $1 - P_{m,i}^{offld}$.

5 s-EPPOS Meta-Heuristic

In [5], we proposed the so-called energy-aware path planning and offload scheduling heuristic (EPPOS) algorithm, which jointly tackles path planning and offload scheduling for drones with energy constraints. Here, we employ a simplified version of the algorithm, referred to as s-EPPOS, assuming that the energy is sufficient for the whole mission. Note that this assumption is also made by the greedy algorithm discussed in the previous section.

The heuristic works in the spirit of a variable neighborhood search (VNS) algorithm [19], gradually improving the best solution found so far by exploring neighboring solutions through random search. In our case, the solution consists of (a) the paths to be followed by the drones and (b) the schedule for offloading their computations to nearby edge servers. In turn, the offloading schedule is produced based on the so-called candidate selection order in which the drones are examined to find the next best possible offloading option. This is encoded as a sequence of drone identifiers, e.g., [1, 2, 3.., 1, 2, 3..] means that the algorithm will first plan the next offloading option for drone d_1, then for d_2, then for d_3, etc. Note that the same identifier may appear more than once in this sequence, each such occurrence dictating when the algorithm will plan the next offload for that drone with respect to others.

The high-level logic of the heuristic is given in Algorithm 2. The algorithm takes as input the set of drones $\mathcal{D}$, an initial path P_m^{init} for each drone d_m (computed using an off-the-shelf TSP algorithm), and the number of iterations of the optimization loop.

As a first step, the candidate selection order is set to a randomly chosen round-robin order and the best minimum relative reduction is initialized.

Then the algorithm iteratively improves the paths and offloading schedule. Each iteration starts by resetting the path P_m of each d_m to P_m^{init} and adjusting these paths. The path adjustment is done in a similar manner as in the first step of the greedy algorithm. The algorithm picks the drones in a random order (different in each iteration) and tries to reverse the path of each drone. If a reversal reduces the contention in the servers, then this reversed path is adopted.

After the paths are reset and adjusted, a new offloading schedule is produced via the SCHEDULEOFFLOADING() function. This starts from a fresh state where all servers are unoccupied at all times and no computation offloading is planned

Algorithm 2 s-EPPOS meta-heuristic

```
function MAIN(𝒟, P_m^def, iterations)
    order ← rndRoundRobinOrder(𝒟)
    bestMinReduction ← −1
    for iterations do
        P ← tryReversePaths(P_m^def)
        S, W, U, reductions ← SCHEDULEOFFLOADING(𝒟, P, order)
        minReduction ← min(reductions)
        if minReduction > bestMinReduction then
            bestMinReduction ← minReduction
            bestP, bestS, bestW, bestU ← P, S, W, U
            bestOrder ← order
        end if
        order ← SHAKEORDER(bestOrder, reductions)
    end for
end function

function SCHEDULEOFFLOADING(𝒟, P, order)
    for each d_m ∈ 𝒟 do
        pos_m ← 1                                         ▷ start from P_m[1] = dep_m
        S_m[i] = 0, W_m[i] = 0, 1 ≤ i ≤ len(P_m)           ▷ no offloading / waiting times yet
        U[k][m][t] ← 0, ∀k, t                              ▷ no server reservations yet
    end for
    while order ≠ ∅ do
        d_m ← pickNxtDrone(order)
        pos_m, k, t^arr, waitT ← fndNxtOffload(pos_m, P_m, S, W, U)
        if pos_m ≠ len(P_m) then                           ▷ offload option found
            S_m[pos_m], W_m[pos_m] ← k, waitT
            U[k][m][t] ← 1, ∀t, t^arr + waitT ≤ t ≤ t^arr + waitT + compT(m, k)
        else                                               ▷ arrived at the end of the path
            order ← rmvAllEntries(order, d_m)
            reductions_m ← (mT_m^def − mT_m) / mT_m^def    ▷ reduction as per Eqs. 4 and 6
        end if
    end while
    return S, W, U, reductions
end function

function SHAKEORDER(order, reductions)
    moves ← pickRndRange(moves^low, moves^high)
    bestD ← sortDescendingReductions(reductions, 𝒟)
    worstD ← sortAscendingReductions(reductions, 𝒟)
    for mfrom1tomoves do
        first_occ ← pickRndRange(first_occ^low, first_occ^high)
        order ← moveFirstEntriesToTail(order, bestD[m], first_occ)
        last_occ ← pickRndRange(last_occ^low, last_occ^high)
        order ← moveLastEntriesToHead(order, worstD[m], last_occ)
    end for
    if pickRndBinary(prob_swap) = true then
        swaps ← pickRndRange(swaps^low, swaps^high)
        for swaps do
            m_1, m_2 ← pickRndPair(order)
            order ← swapEntries(order, m_1, m_2)
        end for
    end if
    return order
end function
```

for any drone. Then, drones are considered according to the candidate selection order, and for each one, the next computation offload is planned. This is done via the $fndNxtOffload()$ function, which finds the next best safe offloading option in the drone's path (not shown for brevity). The function returns the offloading suggestion at the next possible point of interest, along with the respective server and waiting time so that the server capacity constraint as per Eq. 5 is satisfied. If the returned point indeed corresponds to a point of interest, the offloading schedule is updated accordingly. However, if the returned point corresponds to the last point in the drone's path implying that no offloading opportunity was found for d_m, all occurrences of the drone are removed from the ordered candidate selection list.

When the SCHEDULEOFFLOADING() function returns, it is checked whether the newly produced paths and offloading schedule improve the optimization criterion vs the best solution found so far (achieves a higher worst relative reduction of the mission time across all drones). If so, the new paths and schedule, along with the corresponding candidate selection order, are stored as the best solution.

Finally, before entering the next iteration, the best candidate selection order is modified via the SHAKEORDER(). This function changes the best selection order found so far to build a new selection order in which the drones will be considered for offloading in the next iteration. This is done in two different ways. On the one hand, it picks $moves$ drones that achieved the highest relative reduction of their mission times and for each one moves its $first\_occ$ first occurrences in the order list to the tail of the list. Also, it picks the same number of drones that achieved the lowest relative reduction of their mission times and for each one moves its $last\_occ$ last occurrences in the order list to the head of the list. On the other hand, with probability $prob_{swap}$, it swaps the order of two randomly chosen list entries. This is repeated for $swaps$ times. The rationale of the first transformation is to strike a better balance by promoting the drones with the worst mission time improvement so that they are considered earlier for offloading in the next iteration while demoting the drones with the best improvement. The second transformation is a simple random mutation. Note that all the above variables ($moves$, $first\_occ$, $last\_occ$, $swaps$) are randomly picked from specific intervals that are specified via respective low/high bounds.

6 Evaluation

Conducting experiments in the field with a large number of drones is very hard and extremely time consuming. In fact, to be perfectly legal, one would need a standby pilot for each drone. Moreover, it is practically impossible to replicate the exact same conditions in order to compare the outcome of different approaches. Therefore, we evaluate the two algorithms described in Sects. 4 and 5 via simulations. Nevertheless, all key parameters are based on performance numbers that are obtained using real platforms, both for the drone and the edge servers.

We present our evaluation based on two main sets of experiments. First, all of the drones need to visit the same points of interest. Then, subsets of the drone fleet are assigned different points of interest. In all experiments, the drones are heterogeneous both in terms of the onboard platform used to run computations locally and in terms of their flying capabilities.

6.1 *Topology and Missions*

We conduct experiments for a 21 × 21 grid topology, shown in Fig. 1. The nodes represent the potential points of interest (waypoints) for the drones. The horizontal and vertical distance between neighboring nodes is 20 m. All drones start from and land on the same depot node, located in the center of the grid. Further, we assume two edge servers, located at the gray-black-bordered nodes, with an assumed range

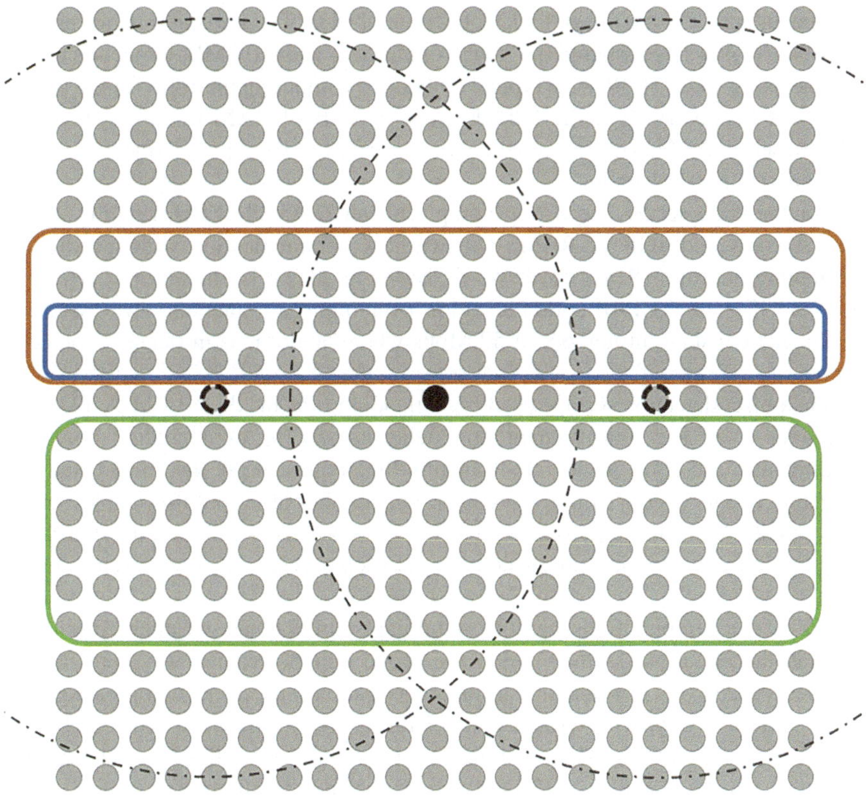

Fig. 1 Topology and missions used in the experiments

of 200 m. The circular regions in the figure mark the regions from where each server is reachable and can be used for offloading.

We experiment with three types of missions, for different regions denoted in Fig. 1 via rounded rectangles: (i) the *small* mission which includes all points in the two rows above the central row of the grid (42 points of interest—blue rectangle), (ii) the *medium* mission which includes the four rows above the central row (84 points of interest—brown rectangle), and (iii) the *large* mission which includes the six rows below the central row (126 points of interest—green rectangle). The default paths provided as input to the algorithms are computed using a conventional TSP algorithm that does not take into account offloading.

6.2 Drone and Server Parameters

We assume a fleet of $M = 24$ drones that need to perform the same sensing and processing task at each point of interest, which is to take a picture using an onboard camera and process the image to detect certain objects. The sensing (image capture) delay $senseT$ is 1 second and the image obtained is about 0.5MB. Object detection is done using YOLO [20]. As a typical embedded computing platform for the drone, we use a Raspberry Pi (RPi) 4 model B with 2GB of RAM running a headless Ubuntu server 20.04. The measured delay $compT_m$ for the required object detection computation is 10 seconds.

For the edge server, we use a laptop with an i7-8550U CPU, running a VM with 4 cores and 4GB of RAM. The wireless communication with the drone's computing platform (RPi) is over Wi-Fi. Based on network simulations performed via NS3 [21] for TCP/IP communication over high gain antennas, we assume a practically constant throughput of 52 Mbps when the drone is in range of the edge server.

The object detection service used for offloading runs inside the VM as a Docker container [22]. The total computation delay $compT(m, k)$, including container loading and the time needed to transmit the request with the image and the detection results over Wi-Fi, is 2 seconds. The number of drones that can be served concurrently by each such server is set to 1. This is because we observed a significant increase in the processing time when several such services run concurrently in the VM.

We use a simple flight model, assuming stable/known weather conditions, constant acceleration 0.4 m/s^2 when the drone departs from a waypoint until it reaches a speed of 4 m/s, and a constant deceleration 0.8 m/s^2 as it approaches and stops at the next waypoint. With this model, the flying time between two neighboring waypoints is about 12 s. The resulting flight behavior is similar to what we have observed via the ArduPilot [23] software-in-the-loop configuration as well as in field tests using a real cheap quadcopter. The default makespan mT_m^{def}, when the drone performs the required computation at all points of interest using its onboard

Table 1 Drone heterogeneity in terms of computing and flight capability

	Type A	Type B	Type C	Type D
Travel speed factor	0.8	0.8	1	1
Local computation factor	0.8	1	0.8	1

Table 2 Settings for the parameters guiding the random search of s-EPPOS

Parameter	Value
$moves^{low}, moves^{high}$	$1, \frac{M}{3}$
$first\_occ^{low}, first\_occ^{high}$	$1, 5$
$last\_occ^{low}, last\_occ^{high}$	$1, 5$
$prob_{swap}$	0.1
$swaps^{low}, swaps^{high}$	$\frac{M}{2}, M$

platform, without any offloading, is roughly 17 min, 33 min, and 49 min for the small, medium, and large missions, respectively.

The heterogeneity between drones is introduced as follows. On the one hand, we let some drones perform the computation 20% faster than what was measured with the RPi, assuming that these drones feature faster computing platform. On the other hand, we assume that some drones feature stronger motors or lighter payload, which allows them to travel faster, covering the same distance in 20% less time than what is estimated in the above flight model. Combining these two heterogeneity aspects, we consider four different types of drones, shown in Table 1. More precisely, drones of types A and C can compute faster, while drones of types A and B can fly faster. Note that type A drones have the best settings in terms of both computation and travel speed, while type D drones have the worst settings.

In all experiments, we assume that all 24 drones operate in the same area and compete for the shared edge resources (as per Fig. 1). Also, for the sake of symmetry, we equally divide drones among the different types A, B, C, and D; in other words, we have six drones of each type. The settings used to guide the random search of the s-EPPOS meta-heuristic are given in Table 2.

6.3 Results: Same Missions

In a first set of experiments, we investigate the scenario where all of the drones need to visit the same points of interest, for the small, medium, and large missions. Figure 2 shows the results in the form of boxplots. In each plot, the first three boxes correspond to the different policies of the greedy algorithm described in Sect. 4, while the fourth boxplot captures the results of the s-EPPOS meta-heuristic. The Y axis denotes the relative reduction vs the default makespan. On each box, we mark the relative reduction of each individual drone using different marker shapes for the four different drone types. Also, on the bottom of each box, we give the worst relative reduction (smallest value among all drones), which is what we wish

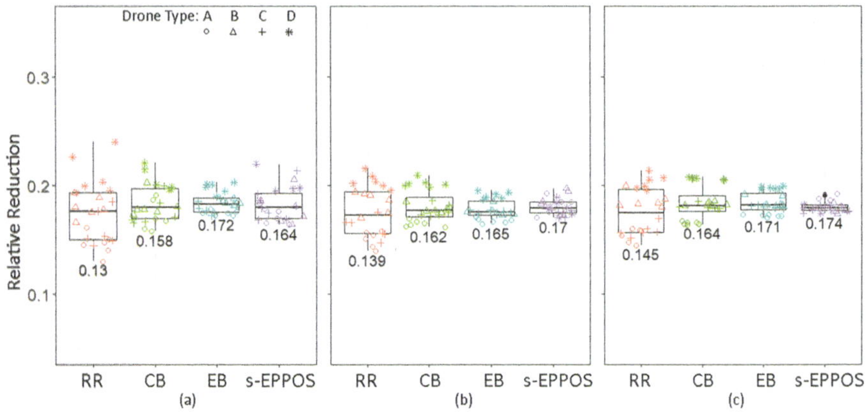

Fig. 2 Same missions: (**a**) small, (**b**) medium, and (**c**) large

to maximize as per Eq. 6. The results for the small, medium, and large missions is shown Fig. 2a–c, respectively.

We observe that the best results are found by the EB policy of the greedy algorithm and the s-EPPOS meta-heuristic. On the one hand, EB considers the contention between the drones and the expected total relative reduction that each drone will achieve, thereby finding a balanced and fair schedule. EB finds schedules with 16.9% average worst relative reduction over all missions, while the average standard deviation between the relative reduction of each drone is 0.95%. On the other hand, s-EPPOS iteratively explores multiple randomized solutions, converging to a good and fair plan. The average worst relative reduction achieved by s-EPPOS is also 16.9%, while the average standard deviation over all missions is 0.94%, practically the same as EB.

The CB greedy policy finds a little worse results than EB, achieving an average worst relative reduction of just 16.1%. The reason is that CB does not consider the contention between drones, making it harder to share the edge resources in a fair way. Due to the heterogeneity among drones, some complete their mission faster than others. This means that some drones will be exposed to less contention in total than others. The lack of fairness can also be observed by the increased standard deviation between the drones which for CB is 1.55% on average over all missions.

Finally, the RR heuristic finds the worst and most unfair schedules vs all other approaches. This is because it blindly gives edge resources to each drone without considering the resulting relative reduction or contention. The worst relative reduction and the standard deviation are 13.8% and 2.41%, respectively, averaged over all missions. Notably, s-EPPOS outperforms RR by 3.4%, 3.1%, and 2.9% for the small, medium, and large missions, respectively.

By looking at the individual drone markers on the boxplots, we can observe that in all the versions of the greedy algorithm (RR, CB, and EB), the worst improvement is achieved for the type A drones. This is reasonable since these drones have the best settings, making any optimizations less important. On the one hand, the shorter local computation time makes the time difference vs offloading smaller. On the other hand, due to the shorter travel times, type A drones are the first to complete their missions, leaving more room (less contention) for the rest of the drones, while they are exposed to the highest contention (with all other drones) during their entire mission. For the opposite reasons, the best results are achieved for type D drones. The type C drones achieve worst improvements compared to types B and D. This is because they have fast hardware, so offloading becomes less significant for the overall mission time.

The s-EPPOS meta-heuristic does not follow this pattern. The reason is the way this algorithm operates. Namely, s-EPPOS finds a good schedule after multiple iterations of randomly shuffling the decision order of the drones. In these mutations, the algorithm does not take into account the type of the drone and (in part) does not even consider which drone actually has the worst expected relative reduction of the mission time. As a consequence of this randomized approach, s-EPPOS does not favor a particular type of drone.

Finally, an interesting observation is that for the small missions, shown in Fig. 2a, EB outperforms s-EPPOS even if slightly by 0.8%. The reason is that EB builds a schedule based on the estimated future relative reduction of all drones. If the estimation is close to the real relative reduction that a drone will achieve, then this approach is more appropriate/accurate than building a solution randomly, even when the random procedure is repeated for a number of iterations. However, this is not the case for medium and larger missions shown in Fig. 2b, c, where s-EPPOS outperforms EB by 0.5% and 0.3%. This is because the estimation of the expected reduction of mission time done by EB becomes less accurate as the length of the mission increase as in these cases the future is less predictable.

6.4 Results: Different Missions

In a second set of experiments, the drones need to visit different points of interest according to the three cases shown in Fig. 1. We explore three scenarios. In the first scenario, half of the drones are assigned the small mission and the other half follow the medium mission. In the second scenario, half of the drones are assigned the small mission and the other half follow the large mission. In the third scenario, half of the drones are assigned the medium mission and the other half follow the large mission. For the sake of symmetry, we divide the drones of each type so that half of them are assigned the smaller mission (small or medium), while the other half follow the larger mission (medium or large). For example, *three* type A drones follow the smaller mission and the other *three* follow the larger mission. The same applies for the rest of the drone types. Figure 3 shows the results for each scenario in the form of boxplots.

First of all, we observe that the standard deviation of the relative reduction achieved by each drone is greater for the RR, CB, and even EB versions of the

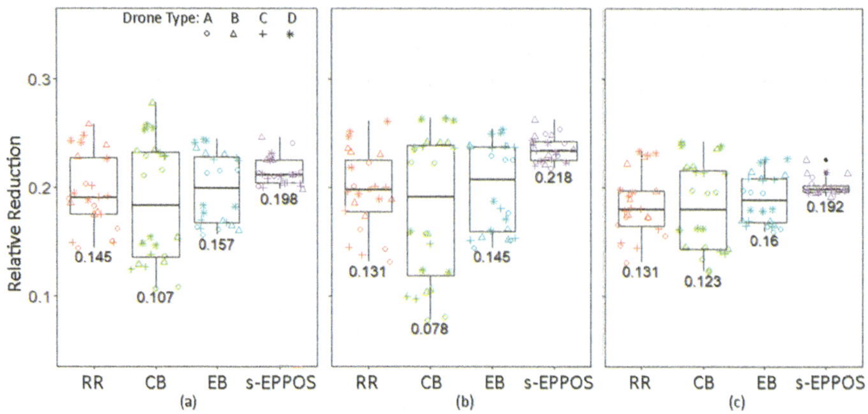

Fig. 3 Different missions: (**a**) small-medium, (**b**) small-large, and (**c**) medium-large

greedy algorithm, compared to the previous set of experiments. More precisely, the deviation becomes 3.33% (vs 0.95%), 5.47% (vs 1.55%), and 3.2% (vs 2.41%) for RR, CB, and EB, respectively.

Interestingly, the worse results are achieved by the CB policy rather than RR (as in the previous experiments). This is because CB does not consider the contention, as already discussed in the previous subsection. The impact becomes even more significant here because the length of the missions differs. Since CB myopically considers the current reduction in the mission time of each drone, it gives priority to drones with longer paths because the time saved from an offloading decision is not as significant as the time saved from an offloading when the path is shorter. This bias exists in all three scenarios, as can be confirmed by checking the points that represent the different drone types. We can see that half of the drones of a given type achieve much better relative reduction in the mission time vs the other half. As a result, CB becomes more unfair than RR.

The EB policy achieves better results vs the other two policies, outperforming RR by 1.2%, 1.4%, and 2.9% for the small-medium, the small-large, and the medium-large scenarios, respectively. Still, looking at the data points for the same drone type, we observe a similar pattern as in CB: there is a separation between half of the drones of each type, with better results being achieved for the drones following the longer vs the shorter path. This is because the estimations of EB regarding the total relative reduction of each drone are more uncertain when the path is longer, underestimating the reduction that can be achieved for the drones with the longer path in the future, after the drones with the shorter path will have finished their missions. As a result, the EB policy is slightly more favorable for the drones with the larger missions.

Notably, the difference between s-EPPOS and EB becomes far more significant in this set of experiments, 4.1%, 7.3%, and 3.2% for the small-medium, the small-large, and the medium-large scenarios, respectively. This is because EB fails to estimate the total relative reduction of each drone accurately, leading to unfair schedules vs those produced by s-EPPOS. Furthermore, s-EPPOS achieves a more balanced improvement for all drones at much smaller standard deviation of 1.1% on average over all scenarios, almost $\frac{1}{3}$ of the deviation of the EB greedy policy. Also, like in the previous experiments, s-EPPOS does not favor a particular type of drone, whereas RR, CB, and EB achieve better improvements for drones of types B and D, for the reasons explained in the previous subsection.

7 Conclusion

In this work, we have focused on the problem of the joint path planning and scheduling of computation offloading. We have evaluated a greedy algorithm with three policy variants and a meta-heuristic aiming at evenly reducing the mission time of all drones, for drones with different computing and flight capabilities under various mission scenarios. Our results show that the s-EPPOS heuristic finds better and more balanced results in the vast majority of scenarios. In addition, s-EPPOS achieves uniform improvement even in scenarios where the drones are assigned widely different missions. Still, the greedy EB policy is shown to be quite promising, and it would be interesting to combine the EB logic with the s-EPPOS heuristic.

References

1. F. Mohammed, A. Idries, N. Mohamed, J. Al-Jaroodi, I. Jawhar, Uavs for smart cities: opportunities and challenges, in *Proceedings of the International Conference on Unmanned Aircraft Systems* (2014), pp. 267–273
2. P.K.R. Maddikunta, S. Hakak, M. Alazab, S. Bhattacharya, T.R. Gadekallu, W.Z. Khan, Q.-V. Pham, Unmanned aerial vehicles in smart agriculture: applications, requirements, and challenges. IEEE Sens. J. **21**(16), 17 608–17 619 (2021)
3. E. Semsch, M. Jakob, D. Pavlicek, M. Pechoucek, Autonomous UAV surveillance in complex urban environments, in *Proceedings of the IEEE/WIC/ACM International Joint Conference on Web Intelligence and Intelligent Agent Technology* (2009), pp. 82–85
4. G. Polychronis, S. Lalis, Planning computation offloading on shared edge infrastructure for multiple drones, in *Proceedings of the IEEE International Workshop on Wireless Sensor, Robot and UAV Networks (WiSARN), IEEE International Conference on Distributed Computing Systems Workshops (ICDCS-W)* (2022), pp. 302–307
5. G. Polychronis, S. Lalis, Joint edge resource allocation and path planning for drones with energy constraints, in *Mobile and Ubiquitous Systems: Computing, Networking and Services (MobiQuitous)* (Springer Nature Switzerland, Basel, 2023), pp. 378–399
6. T. Kasidakis, G. Polychronis, M. Koutsoubelias, S. Lalis, Reducing the mission time of drone applications through location-aware edge computing, in *Proceedings of the IEEE International Conference on Fog and Edge Computing (ICFEC)* (2021), pp. 45–52
7. B. Kim, H. Min, J. Heo, J. Jung, Dynamic computation offloading scheme for drone-based surveillance systems. Sensors **18**(9), 2982 (2018)
8. C. Zhan, H. Hu, X. Sui, Z. Liu, J. Wang, H. Wang, Joint resource allocation and 3d aerial trajectory design for video streaming in UAV communication systems. IEEE Trans. Circuits Syst. Video Technol. **31**(8), 3227–3241 (2020)

9. J. Xiong, H. Guo, J. Liu, Task offloading in uav-aided edge computing: bit allocation and trajectory optimization. IEEE Commun. Lett. **23**(3), 538–541 (2019)
10. J. Zhang, L. Zhou, Q. Tang, E.C.-H. Ngai, X. Hu, H. Zhao, J. Wei, Stochastic computation offloading and trajectory scheduling for UAV-assisted mobile edge computing. IEEE Internet Things J. **6**(2), 3688–3699 (2018)
11. A. Al-Hilo, M. Samir, C. Assi, S. Sharafeddine, D. Ebrahimi, UAV-assisted content delivery in intelligent transportation systems-joint trajectory planning and cache management. IEEE Trans. Intell. Transp. Syst. **22**(8), 5155–5167 (2020)
12. M. Asim, W.K. Mashwani, S.B. Belhaouari, S. Hassan, A novel genetic trajectory planning algorithm with variable population size for multi-UAV-assisted mobile edge computing system. IEEE Access **9**, 125 569–125 579 (2021)
13. A. Paulos, S. Dasgupta, J. Beal, Y. Mo, J. Schewe, A. Wald, P. Pal, R. Schantz, J.B. Lyles, Priority-enabled load balancing for dispersed computing, in *Proceedings of the IEEE International Conference on Fog and Edge Computing (ICFEC)* (2021), pp. 1–8
14. G.L. Stavrinides, H.D. Karatza, A hybrid approach to scheduling real-time IoT workflows in fog and cloud environments. Multimedia Tools Appl. **78**(17), 24 639–24 655 (2019)
15. G.L. Stavrinides, H.D. Karatza, Orchestrating real-time IoT workflows in a fog computing environment utilizing partial computations with end-to-end error propagation. Cluster Comput. **24**(4), 3629–3650 (2021)
16. I. Stypsanelli, O. Brun, B.J. Prabhu, Performance evaluation of some adaptive task allocation algorithms for fog networks, in *Proceedings of the IEEE International Conference on Fog and Edge Computing (ICFEC)* (2021), pp. 84–88
17. G.L. Stavrinides, H.D. Karatza, Orchestrating bag-of-tasks applications with dynamically spawned tasks in a distributed environment, in *Proceedings of the International Symposium on Performance Evaluation of Computer and Telecommunication Systems (SPECTS)* (2021), pp. 1–8
18. G. Polychronis, S. Lalis, Flexible computation offloading at the edge for autonomous drones with uncertain flight times, in *Proceedings of the International Conference on Distributed Computing in Smart Systems and the Internet of Things (DCOSS-IoT)* (2023). To appear
19. N. Mladenović, P. Hansen, Variable neighborhood search. Comput. Oper. Res. **24**(11), 1097–1100 (1997)
20. J. Redmon, A. Farhadi, Yolo v3: an incremental improvement (2018). Preprint. arXiv:1804.02767
21. G.F. Riley, T.R. Henderson, The ns-3 network simulator, in *Modeling and Tools for Network Simulation* (2010), pp. 15–34
22. D. Merkel et al., Docker: lightweight linux containers for consistent development and deployment. Linux J. **239**(2), 2 (2014)
23. Ardupilot web site. https://ardupilot.org

Multi-objective Optimization Approach to High-Performance Cloudlet Deployment and Task Offloading in Mobile Edge Computing

Xiaojian Zhu and MengChu Zhou

1 Introduction

With the development of the Internet of Things (IoT) technology, more and more smart devices are used, such as mobile phones, tablets, intelligent sensors, wearable devices, and smart cameras. More and more applications are run on them. If the applications are run on the remote central cloud, the task response delay of users tends to be large as a result of likely high network latency caused by, e.g., network congestion [2, 3, 33, 37]. To overcome this issue, mobile edge computing technology [1, 4, 37, 60, 63, 64] emerges. Computational resources can be brought close to users via a mobile edge computing system (MECS) [1, 37, 41]. It consists of a number of cloudlets, and its deployment can decrease task response delay and network traffic. It is thus regarded as one of the key technologies to cope with the challenges posed by the proliferation of mobile devices and IoT applications [25]. In smart city applications, cloudlets can be co-located with some base stations to process huge amounts of data produced by IoT devices. In addition, cloudlets can be deployed at some important sites, such as railway stations, factories, shopping arcades, schools, and roadside units, to provide computing resources for intelligent terminal devices/equipment. In these applications, it is necessary to optimize energy consumption and task response delay of users and the number of deployed cloudlets. Some of mobile applications are computation-intensive, such

X. Zhu (✉)
School of Computer Science and Engineering, Changshu Institute of Technology, Changshu, Jiangsu, China
e-mail: zhuxj5318@163.com

M. Zhou
Helen and John C. Hartmann Department of Electrical and Computer Engineering, New Jersey Institute of Technology, Newark, NJ, USA
e-mail: zhou@njit.edu

C. Savaglio et al. (eds.), *Device-Edge-Cloud Continuum*, Internet of Things,
https://doi.org/10.1007/978-3-031-42194-5_6

as augmented reality, video processing, and natural language processing. Since they require more computational resources, their executions consume more energy. Because user devices are powered by batteries, an important consideration is how to save their energy [34, 38]. In MECS, user devices can reduce their energy consumption and extend their service life through offloading their nonurgent tasks to cloudlets.

Decreasing response latency of mobile applications improves user experience [28, 46]. Compared to local execution, the response delay of user tasks executed on cloudlets may be longer due to more waiting time including transmission, routing, and execution time. The large number of users sharing limited wireless bandwidth may cause long transmission time from them to their associated access point (AP) or base station (BS). The routing time from an AP to a cloudlet is long if their distance is large. When too many tasks are executed on cloudlets, execution time may become too long due to the cloudlets' limited computing capacities. Therefore, a good trade-off between energy consumption and task response delay of users has to be made.

In addition to task offloading strategies, the optimal deployment of cloudlets according to user demand distribution is also essential to improve the quality of experience (QoE) of users. More cloudlets can accommodate more offloaded tasks of users, and thus more energy of users can be saved. The routing and running time of user tasks can be reduced by deploying more cloudlets at more appropriate sites. However, a larger number of deployed cloudlets lead to higher deployment cost. In order to reduce such deployment cost, the providers of edge computing services strive to deploy as few cloudlets as possible while guaranteeing user QoE. Accordingly, it is necessary to balance MECS deployment cost and user QoE.

Cloudlet deployment has to consider the distribution of application tasks and their offloading in order to meet user QoE and device energy consumption requirements. In other words, the cloudlet deployment and task offloading should be jointly optimized. Different cloudlet deployment schemes incur different task offloading schemes. For each cloudlet deployment scheme, we need to compute the optimal task offloading scheme so as to obtain the maximum gain of this deployment scheme.

Most existing work on cloudlet deployment such as [13, 30, 61] and [44] mainly focuses on the minimization of deployment cost or task response delay, but ignores the optimization of user energy consumption. Although there is some work studying task offloading strategies to minimize energy consumption and task response delay of users, e.g., [22, 32, 53], it does not take into account cloudlet deployment cost. On the other hand, most existing work on cloudlet deployment and task offloading such as [12, 13, 22, 53] and [62] employs the weighted sum method to deal with multi-objective optimization problems, while it is hard to select such weighting coefficients. Observing the limitations of the existing work, we are motivated to examine how to optimize energy consumption and task response delay of users and MECS deployment cost simultaneously, such that a comprehensive insight into MECS design can be offered to MECS service providers. In addition, we are motivated to develop an effective method for producing a set of trade-off solutions, such that it is easy for them to make an informed decision. Specially, an MECS service provider can choose an appropriate cloudlet deployment scheme from the set of nondominated solutions delivered by our proposed method to make desired trade-off among all the three objectives.

In this work, we investigate a multi-objective optimization problem arising from cloudlet deployment and task offloading (MOCT) in MECS. In this problem, not only the minimization of energy consumption and task completion delay of users is required, but also the minimization of network deployment cost is demanded. The three objectives conflict with each other. Specifically, less energy consumption of users makes more tasks offloaded, and thus more cloudlets have to be deployed. On the other hand, smaller task completion delay of offloaded tasks requires less routing and running time, thus increasing the number of deployed cloudlets. Moreover, lower task response latency allows fewer tasks to be offloaded, yet increasing the energy consumption of user devices. As the three objectives are essentially very different, their weighted sum makes little sense to a decision-maker. Therefore, we do not use weighting coefficients to convert MOCT into a single-objective optimization problem. Instead, a set of Pareto optimal solutions have to be found. This work attempts to make the below contributions:

1. Formulating a novel problem of MOCT as a mixed integer nonlinear program (MINP) and proving it to be NP-complete.
2. Proposing a modified guided population archive whale optimization algorithm for MOCT to produce a set of Pareto optimal solutions given an MINP. In addition, this work conducts extensive simulations to compare the performance of our proposed algorithm against those in the literature. The comparison results confirm the superiority of our proposed algorithm over its compared peers.

The organization of the rest of this chapter is as follows. Section 2 reviews related work on mobile edge computing. Section 3 builds a mathematical model for MOCT and analyzes its complexity. Section 4 gives a modified guided population archive whale optimizer-based cloudlet deployment and task offloading algorithm. Section 5 reports the simulation results. Section 6 summarizes the work.

2 Overview of Related Work

An MECS can provide computational resources for users. On the one hand, to reduce energy consumption and task completion delay of users, the problems of task offloading and resource allocation should be effectively solved. On the other hand, the cloudlets should be properly deployed so as to enhance system performance while minimizing system cost.

During task offloading, communication and computation resources have to be optimally allocated subject to the resource and quality-of-service (QoS) limitations. The work in [22] focuses on how to optimize the computing frequencies and transmission powers of IoT mobile devices and make offloading decisions of their multiple types of tasks to minimize their task completion delay and energy consumption under the maximum completion latency constraint of each task. Tran and Pompili [53] investigate how to optimally select a task offloading strategy, user transmission power, and allocation of computing frequencies of mobile edge computing servers to their associated users so as to maximize the system utility with respect to the task completion delay and energy consumption of users. The work in [12] stresses the minimization of the weighted sum of the energy consumed by users for offloading their tasks to cloudlets and the cost of the computational resources of cloudlets allocated to users by network operators while meeting users' maximum task completion delays. Three algorithms are presented to solve the problem. Liu et al. [35] investigate a task offloading problem where offloading probabilities and transmission powers of users are optimized to minimize their energy consumption, task completion latency, and the cost of the resources of the fog node and central cloud used by them. They propose an algorithm based on the interior point method to solve the problem. However, only one fog node is considered in the problem. Khalili et al. [32] consider the problem of minimizing the energy consumption of users by optimizing their offloading decisions and the assignment of subchannels and transmission powers used by them while guaranteeing their maximum service delay constraint. In [7], a modified nondominated sorting genetic algorithm II (NSGA-II) is proposed to tackle the task offloading decision problem for the minimization of both energy consumption and task completion latency of smart mobile devices. Hu et al. [26] employ an unmanned aerial vehicle (UAV) to provide computation resources for ground users. They investigate how to optimize a UAV trajectory and user task offloading to minimize task delay. This problem takes into account the user fairness and the energy consumption constraints of both users and UAV. The work in [14, 15, 17] investigates workload allocation and balancing to minimize task response time of users. However, these studies fail to consider cloudlet deployment.

Coupled with the schemes of task offloading and resource allocation, a cloudlet deployment strategy has a great influence on system performance and cost. Wang et al. [56] examine the problem of deploying edge servers and assigning BSs to edge servers in order to minimize the maximum workload difference among edge servers and the maximum access delay between BSs and edge servers. To solve this problem, a mixed integer program is presented. Xu et al. [61] address how to place cloudlets and assign user requests to minimize the average access delay from users to cloudlets. In this problem, different cloudlets may have different computing capacities, and different user requests may have different computational resource demands. They propose an integer linear program and a heuristic algorithm to solve the problem. In addition, two approximation algorithms are designed for a special case of the problem. The work in [62] reports the cloudlet deployment and task assignment problem with the minimum energy consumption and proves it to be NP-hard. In the problem, the total energy consumption consists of the energy consumed by the used links, placed cloudlets, and cloud servers, and the task completion delay requirement must be met. To formulate and solve the problem, a mixed integer linear program and an algorithm based on the Benders decomposition are presented, respectively. Jia et al. [30] investigate the cloudlet placement and user assignment problem with the minimum system response delay and devise two heuristic algorithms to solve it. In [13], a mixed integer program is proposed for the cloudlet deployment problem with the objective of minimizing the sum of the deployment cost and the delay from the regions of user requests to their assigned cloudlets. In [16], the cloudlets are deployed through a Lagrangian heuristic algorithm. The work in [42] presents a mixed integer linear program for the minimum cost deployment of the roadside units equipped with a server under the constraints of the minimum ratios of the network coverage and satisfied computational resource demand. The deployment cost is related to the number of the deployed roadside units, the distances between them and the cells they cover, and the transmission power levels assigned to them. Ren et al. [44] examine the problem of deploying the minimum number of cloudlets to satisfy the computation demand of each wireless access point subject to the maximum access delay constraint. They also present an integer linear program and a greedy algorithm to solve it. However, these cloudlet deployment strategies fail to simultaneously optimize energy consumption and task response delay of users and MECS deployment cost.

Different from the previous work, we address the problem of the joint optimization of cloudlet deployment and task offloading to minimize the energy consumption and task completion delay of users and the number of deployed cloudlets simultaneously in this work. Because it is a triple-objective optimization problem, we need to devise an effective algorithm to obtain a set of Pareto optimal solutions.

3 Problem Model

We consider a mobile edge computing network with M users and N APs. A number of cloudlets are deployed in the network to provide computing service for users. All APs are interconnected by wired links and each user can access the network via its associated AP. Let u_i $(1 \leq i \leq M)$ and v_j $(1 \leq j \leq N)$ represent the ith user and jth AP, respectively. Denote ϑ_i and λ_i as the index of the associated AP and average task arrival rate of u_i, respectively. Denote L as the maximum number of the cloudlets that can be deployed, which is the upper bound of the budget of a service provider. Let c_i stand for the ith $(1 \leq i \leq L)$ cloudlet. We assume that the potential sites where the cloudlets can be deployed are the positions of the APs, i.e., the deployed cloudlets are co-located with some of the APs. In addition, at most a cloudlet can be deployed at the location of an AP. Figure 1 shows the MECS model we consider. A task of a user u_i can be characterized by its input data size and computational resource demand. Both of them are assumed to satisfy the exponential distribution, and their average values are represented by α_i (bits) and β_i (CPU cycles), respectively. Denote φ_i^u (CPU cycles/second) and ξ_i as the computing capacity and effective switched capacitance of the CPU of u_i, respectively.

The energy consumption model for processing a task in [22, 29, 53] is adopted, and thus the energy consumption for executing a task of a user u_i locally is given as

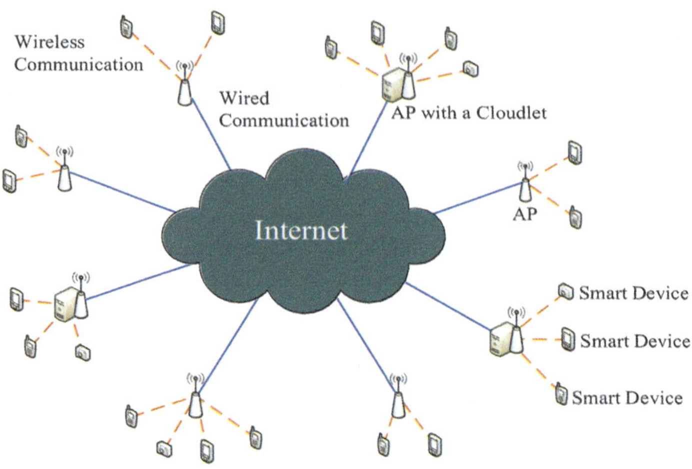

Fig. 1 Illustration of an MECS model

$$e_i^c = \xi_i (\varphi_i^u)^2 \beta_i. \tag{1}$$

As shown in Fig. 1, if a user wants to offload some of its tasks to some cloudlets, it has to first transmit them to its associated AP through the wireless link. Then, its associated AP sends them to the APs where the cloudlets are located through wired links. In the network, the orthogonal frequency-division multiple access (OFDMA) is adopted. All users share the same system bandwidth B, and the users associated with the same AP employ orthogonal channels to transmit their offloaded tasks. Thus, there is no interference among users associated with the same AP, but users associated with different APs may interfere with each other. Let p_i and w_i stand for the transmission power and available channel bandwidth of user u_i, respectively. For the users associated with the same AP, the system bandwidth is evenly divided among them, and thus $w_i = B/\sum_{1\le j\le M \wedge \vartheta_j=\vartheta_i} 1$. Denote $g_{i,j}$ and σ^2 as the channel gain from u_i to v_j and the background noise power, respectively. The transmission rate of u_i can be computed as [10]

$$r_i^u = w_i log_2 \left(1 + \frac{p_i g_{i,\vartheta_i}}{\sigma^2 + \sum_{1\le j\le M \wedge \vartheta_j \ne \vartheta_i} p_j g_{j,\vartheta_i}}\right). \tag{2}$$

The energy consumption required for transmitting a task of a user u_i to its associated AP is

$$e_i^t = p_i \frac{\alpha_i}{r_i^u}. \tag{3}$$

Denote $z_{i,j}$ as the probability of offloading a task of user u_i to cloudlet c_j. It is assumed that the task arrival at a user is a Poisson process, and thus both the local execution and transmission of its tasks are modeled as an M/M/1 queue [58]. Accordingly, the average waiting time composed of the queue and execution time for executing a task of u_i locally is

$$\tau_i^l = \frac{\beta_i}{\varphi_i^u - \beta_i\lambda_i(1 - \sum_{1\le j\le L} z_{i,j})}. \tag{4}$$

The average waiting time consisting of the queue and transmission time for transmitting an offloaded task of u_i to its associated AP is

$$\tau_i^t = \frac{\alpha_i}{r_i^u - \alpha_i\lambda_i \sum_{1\le j\le L} z_{i,j}}. \tag{5}$$

Each cloudlet is equipped with only one server. Denote φ^c and μ as the computing capacity (CPU cycles/second) and maximum workload (CPU cycles/second) of each cloudlet, respectively. The maximum workload of a cloudlet is leveraged to avoid the extremely long waiting time of offloaded tasks. Since the amount of the resulting data of a task is very small, we ignore the energy consumption and delay caused by returning it from a cloudlet to a user. Since the task arrival at a cloudlet is a Poisson process, the task execution in it can be modeled as an M/M/1 queue [58]. Thus, the average waiting time comprising of the queue and execution time for executing a task in a cloudlet c_i is

$$\tau_i^c = \frac{\frac{\sum_{1\le j\le M} z_{j,i}\lambda_j\beta_j}{\sum_{1\le j\le M} z_{j,i}\lambda_j}}{\varphi^c - \sum_{1\le j\le M} z_{j,i}\lambda_j\beta_j}. \tag{6}$$

Denote $r_{i,j}^v$ as the transmission rate of a link $\langle v_i, v_j\rangle$. We model the connectivity of all APs as a graph G, whose vertex set and arc set are the set of all APs and the set of all links between APs, respectively. The weight of an arc $\langle v_i, v_j\rangle$ in G is the transmission delay of a bit data, i.e., $1/r_{i,j}^v$. Denote $\tau_{i,j}^p$ as the communication delay of a bit data from v_i to v_j, which is the length of the shortest path from v_i to v_j in G.

The total energy consumption of a user u_i is given as

$$e_i = e_i^c\lambda_i\left(1 - \sum_{1\le j\le L} z_{i,j}\right) + e_i^t\lambda_i \sum_{1\le j\le L} z_{i,j}. \tag{7}$$

Denote χ_i as the index of an AP where a cloudlet c_i is deployed. The average response time of a task of user u_i is

$$\tau_i = (1 - \sum_{1\le j\le L} z_{i,j})\tau_i^l + \sum_{1\le j\le L} z_{i,j}(\tau_i^t + \alpha_i\tau_{\vartheta_i,\chi_j}^p + \tau_j^c). \tag{8}$$

Before formulating MOCT, we explain some concepts related to multi-objective optimization problems [6, 18, 21, 59]. A general form of multi-objective optimization problems can be described as

$$\min \quad (f_1(Q), f_2(Q), \ldots, f_m(Q))^{\mathrm{T}} \tag{9}$$

subject to

$$Q \in \Omega \tag{10}$$

where $Q = (q_1, q_2, \ldots, q_n)^{\mathrm{T}}$ is the decision variable vector and Ω is the feasible solution space. A solution Q_1 dominates another solution Q_2 if $(f_1(Q_1), f_2(Q_1), \ldots, f_m(Q_1))^{\mathrm{T}} \neq (f_1(Q_2), f_2(Q_2), \ldots, f_m(Q_2))^{\mathrm{T}}$ and for each objective function f_i $(1 \leq i \leq m)$ $f_i(Q_1) \leq f_i(Q_2)$ [6, 21]. A solution is Pareto optimal if it cannot be dominated by any other one. All Pareto optimal solutions and their objective vectors constitute the Pareto set and the Pareto front, respectively [6, 21].

The MOCT problem aims to select the optimal sites to deploy cloudlets and probabilities to offload tasks of users to cloudlets such that the average energy consumption, task response delay of users, and the number of deployed cloudlets are minimized. Let $\chi'_{i,j}$ denote a binary variable that is 1 if c_i is deployed at v_j, and 0 otherwise. We formulate MOCT as the following MINP:

$$\min \ \psi_1 = \frac{1}{M} \sum_{1 \leq i \leq M} e_i \tag{11}$$

$$\min \ \psi_2 = \frac{1}{M} \sum_{1 \leq i \leq M} \tau_i \tag{12}$$

$$\min \ \psi_3 = \sum_{1 \leq i \leq L} \sum_{1 \leq j \leq N} \chi'_{i,j} \tag{13}$$

subject to

$$\chi'_{i,j} \in \{0, 1\}, \forall 1 \leq i \leq L, \forall 1 \leq j \leq N \tag{14}$$

$$0 \leq z_{i,j} \leq 1, \forall 1 \leq i \leq M, \forall 1 \leq j \leq L \tag{15}$$

$$\sum_{1 \leq j \leq N} \chi'_{i,j} \leq 1, \forall 1 \leq i \leq L \tag{16}$$

$$\sum_{1 \le i \le L} \chi'_{i,j} \le 1, \forall 1 \le j \le N \tag{17}$$

$$\sum_{1 \le j \le L} z_{i,j} \le 1, \forall 1 \le i \le M \tag{18}$$

$$\sum_{1 \le j \le M} z_{j,i} \lambda_j \beta_j \le \mu, \forall 1 \le i \le L. \tag{19}$$

The objective functions (11) and (12) minimize the average energy consumption of all users and the average response delay of all tasks of all users, respectively, while the objective function (13) minimizes the number of deployed cloudlets. Since the energy and waiting time of users are very limited, it is very necessary to reduce their energy consumption and waiting time, which is embodied in (11) and (12). Constraints (14) and (15) define the domains of decision variables. Inequalities (16) and (17) guarantee that a cloudlet is deployed at no more than one site and that at most one cloudlet is deployed at each AP, respectively. Constraint (18) ensures that the total task offloading probability of each user is no more than 1. The maximum workload constraint of each cloudlet is expressed by inequality (19). The important notations used in the problem model are listed in Table 1.

Theorem 1 *The MOCT problem is NP-complete.*

Proof In order to prove the NP-completeness of MOCT, we first show that it belongs to the NP class and then reveal its NP-hardness.

Given a solution to MOCT, it takes $O(ML)$ time to verify both whether the sum of the task offloading probabilities of each user is no more than 1 and the maximum workload constraint of each cloudlet. In addition, it consumes $O(L)$ time to check whether the number of the cloudlets deployed at each site is no larger than one. Thus, to verify a solution to MOCT requires polynomial time. Therefore, MOCT belongs to the class of NP problems.

The NP-hardness of MOCT is proved by showing that the NP-hard p-median problem in [45] is a special case of MOCT. The p-median problem in [45] is to select p facilities from m $(m > p)$ potential ones into a subset such that the sum of the distances from users to their nearest facilities in the subset is minimized.

When $\forall 1 \le i \le M, \varphi_i^u = 0$, $\varphi^c = \infty$, and the price of each cloudlet is 0, the energy consumption of users is fixed, the network deployment cost can be ignored, and all L cloudlets have to be deployed to minimize task response delay of users. Then, MOCT becomes the problem of selecting L APs from N ones to deploy L cloudlets and assigning each user to a cloudlet such that the sum of the communication delays from all users to the cloudlets to which they are assigned is minimized. To this end, each user has to be assigned to a cloudlet to which the communication delay from it is the shortest. In this situation, MOCT becomes the p-median problem in [45]. Because the p-median problem is NP-hard [45], so is MOCT. Hence, MOCT is NP-complete.

Table 1 Important notations

Notation	Meaning
M	Number of users
N	Number of APs
u_i	User i
v_i	AP i
ϑ_i	Index of the associated AP of u_i
λ_i	Average task arrival rate of u_i
L	Maximum number of the cloudlets that can be deployed
c_i	Cloudlet i
α_i	Average input data size of a task of u_i (bits)
β_i	Average computational resource demand of a task of u_i (CPU cycles)
φ_i^u	Computing capacity of u_i (CPU cycles/second)
ξ_i	Effective switched capacitance of the CPU of u_i
e_i^c	Energy consumed by u_i for the local execution of its a task
p_i	Transmission power of u_i
B	System bandwidth
w_i	Available wireless channel bandwidth of u_i
$g_{i,j}$	Channel gain from u_i to v_j
σ^2	Background noise power
r_i^u	Transmission rate of u_i
e_i^t	Energy consumption required for transmitting a task of u_i to its associated AP
$z_{i,j}$	Ratio of the tasks of u_i offloaded to c_j
τ_i^l	Average waiting time for executing a task of u_i locally
τ_i^t	Average waiting time for transmitting a task of u_i to its associated AP
φ^c	Computing capacity of each cloudlet (CPU cycles/second)
μ	Maximum workload of each cloudlet (CPU cycles/second)
τ_i^c	Average waiting time for executing a task in c_i
$r_{i,j}^v$	Transmission rate of a link $\langle v_i, v_j \rangle$
$\tau_{i,j}^p$	Communication delay of a bit data from v_i to v_j
e_i	Total energy consumption of u_i
χ_i	Index of the AP where c_i is deployed
$\chi'_{i,j}$	A binary variable that is 1 if c_i is deployed at v_j, and 0 otherwise
τ_i	Average response time of a task of u_i

4 Heuristic Algorithm

For an MECS service provider, it is hard to determine the constraints. If the constraints are inappropriate, no feasible solution can be found. If we address a constrained single-objective optimization problem rather than MOCT, only an objective is emphasized and only a solution can be generated, which may not make a satisfactory trade-off among all the three objectives. By solving MOCT, a set of nondominated solutions can be produced. Therefore, a comprehensive insight into

MECS design can be offered to MECS service providers, and thus it is highly useful for them to make an informed decision. Furthermore, no matter which objectives need to be stressed or what constraints are considered, the providers can select an appropriate solution from the set of nondominated solutions delivered by our proposed method.

Since MOCT is an NP-complete triple-objective optimization problem and also an MINP problem, it is infeasible for any exact solution methods to solve MOCT in a reasonable time by using scientific computing software, e.g., CPLEX. Therefore, an effective heuristic algorithm should be developed to solve MOCT. Fortunately, meta-heuristic algorithms are very powerful tools that can be adopted to solve MOCT. In this section, we propose a modified guided population archive whale optimization-based cloudlet deployment and task offloading (called MGW for short) algorithm for MOCT.

4.1 Guided Population Archive Whale Optimization Algorithm

The whale optimization algorithm (WOA) [39] models the predatory behavior of a humpback whale swarm and is a powerful tool for solving complex multidimensional single-objective optimization problems. In WOA [39], the positions of a number of whales are iteratively updated to search for a global optimal solution. There are three methods of updating a whale position, i.e., the prey encircling, spiral movement, and prey searching [39]. Denote H and K as the whale population size and the number of components of each whale position, respectively. Let $S_i^t = (s_{i,1}^t, s_{i,2}^t, \ldots, s_{i,K}^t)$ represent the position of the ith whale in the tth generation. Denote S^* as the best solution found so far. To update whale positions using the prey encircling and prey search, two factors $\delta = a(2\gamma - 1)$ and $\zeta = \varpi\gamma$ have to be first computed, where a is the distance coefficient, ϖ is the prey coefficient, and γ is a uniformly distributed random real number in the range of [0, 1]. If the prey encircling is used, the position of whale i is updated as [39]

$$S_i^{t+1} = S^* - \delta|\zeta S^* - S_i^t|. \tag{20}$$

If the spiral movement is selected, the position of whale i is updated as [39]

$$S_i^{t+1} = |S^* - S_i^t|e^{bl}cos(2\pi l) + S^*, \tag{21}$$

where b is the logarithmic spiral shape parameter and l is an evenly distributed random real number in the range of $[-1, 1]$. If the prey search is adopted, the position of whale i is updated as [39]

$$S_i^{t+1} = S_{rand} - \delta|\zeta S_{rand} - S_i^t|, \tag{22}$$

where S_{rand} is the current position of a randomly selected whale.

To handle multi-objective optimization problems, a guided population archive whale optimization algorithm (GPAWOA) is presented in [21]. GPAWOA [21] extends WOA to deal with the problems with more than one optimization objective by making use of an external archive with a fixed capacity and the crowded-comparison operator in the nondominated sorting genetic algorithm II (NSGA-II) [11] to keep nondominated solutions. Specially, in each generation, after the positions of all whales are updated, the newly found nondominated solutions are used to update the archive. If the number of the nondominated solutions in the archive exceeds its capacity, those with the smallest crowding distances are removed. The method of computing the crowding distances is proposed in [11]. The crowding distance of a solution in a nondominated set is the sum of its elementary distances with respect to all objective functions. For each objective function, all solutions in the nondominated set have to be sorted by the value of this objective function first. For a solution, if the value of this objective function is maximum or minimum, its elementary distance regarding this objective function is set to the infinity, otherwise the absolute value of the difference between the normalized values of this objective function at its previous and next neighbors [11]. If the prey encircling or spiral movement is adopted to update a whale position, a leader S^* needs to be selected from the archive to guide the update.

4.2 Whale Position Representation

A position of whale i, i.e., S_i, represents a solution for MOCT and is defined by a 3-tuple (X, Y, Z), where X is an L-vector of indexes of deployment sites of cloudlets, Y is an M-vector of workloads of users, and Z is an $M \times L$ matrix of task offloading probabilities of users. An element $S_i.X[j]$ of $S_i.X$ represents the index of the deployment site of a cloudlet c_j, and its lower bound is set to $2 - N$ so that c_j has the equal opportunities to be deployed and not deployed. If $S_i.X[j] \le 0$, c_j is not deployed. An element $S_i.Y[j]$ of $S_i.Y$ denotes the workload of user u_j. An element $S_i.Z[j][k]$ of $S_i.Z$ stands for the probability of offloading a task of user u_j to cloudlet c_k.

During a search process, in a new position of whale i, $\lambda_j - S_i.Y[j]$ gives the upper bound of the total task offloading flow of user u_j. If only the task offloading probabilities of each user are updated, its total offloaded task flow is easy to be large, and thus its workload is easy to be small, which results in poor convergence and diversity of the obtained nondominated solution set. By means of updating both the workloads and the task offloading probabilities of users, the optimization ability of the proposed algorithm can be enhanced. Each component of $S_i.X$ is rounded to its nearest integer during the whale position updating. Under this representation, the length of a whale position is $K = L + M(L + 1)$.

4.3 Whale Position Initialization

In order to improve the convergence rate, each whale position is initialized with a feasible solution, which is produced by a random feasible solution construction algorithm (RFSCA). In RFSCA, after the number of the cloudlets to be deployed is randomly generated, the allocation of offloaded tasks of users is performed. For each user, RFSCA first randomly generates its total offloaded task flow and then randomly allocates it to some cloudlets. During this allocation, it should be ensured that the workload of each cloudlet is not greater than its upper bound. At last, the cloudlets that have some workload are randomly deployed at some APs. During this deployment, it has to be guaranteed that at most one cloudlet is placed at an AP. In RFSCA, functions $rand()$ and $rint(i, j)$ generate an evenly distributed random real number in [0, 1] and a uniformly distributed random integer in $\{i, i+1, \ldots, j\}$, respectively.

RFSCA is illustrated in Algorithm 1. Herein, lines 1–2 initialize the solution to be constructed. The number of the cloudlets to be placed is randomly produced at line 3. The while-loop at line 7 carries out the allocation of offloaded task flows to cloudlets for users until there is no usable cloudlet. The total offloaded task flow of a user is randomly generated at line 9. The while-loop at line 12 executes the offloaded task flow assignment for a single user, and the remainder is assigned by lines 24–34. A usable cloudlet is randomly chosen at line 13, and lines 15 and 16 randomly generate the offloaded task flow increment, whose computational resource demand cannot be more than the residual available computing capacity of the chosen cloudlet. Lines 17–19 update the task offloading probability matrix, residual offloaded task flow to be allocated, and workload of the chosen cloudlet, respectively. After the task offloading probabilities of all users are determined, their workloads are computed at line 36. The for-loop at line 39 deploys all the used cloudlets. Line 40 deploys an employed cloudlet at a randomly selected usable AP, and line 41 eliminates the AP where the cloudlet has been placed.

In RFSCA, the execution time of lines 1–6 is $O(ML)$. Since the while-loop at line 12 requires $O(L)$ time, it takes $O(ML)$ time to execute the while-loop at line 7. The for-loop at line 39 consumes $O(L)$ time. Therefore, the time complexity of RFSCA is $O(2ML + L) = O(ML)$.

4.4 Whale Position Repairing

In an invalid whale position, more than one cloudlet is deployed at an AP, the constraint of the maximum workload of each cloudlet is violated, or the sum of the offloaded task flows from a user to cloudlets exceeds the difference between the task arrival rate and workload of the user. During the search process of MGW, the invalid whale positions are repaired so as to speed up its convergence. During whale position repairing, the task offloading probabilities of users are first adjusted under

Algorithm 1 RFSCA

Input: $u_1, u_2, \ldots, u_M$; $v_1, v_2, \ldots, v_N$; $c_1, c_2, \ldots, c_L$; ϑ_1, ϑ_2, ..., ϑ_M; λ_1, λ_2, ..., λ_M; $\alpha_1, \alpha_2, \ldots, \alpha_M$; $\beta_1, \beta_2, \ldots, \beta_M$; $\varphi_1^u, \varphi_2^u, \ldots, \varphi_M^u$; B; φ^c; μ
Output: whale position S

$\forall 1 \le i \le L, S.X[i] \leftarrow 0;$
$\forall 1 \le i \le M, \forall 1 \le j \le L, S.Z[i][j] \leftarrow 0;$
$\eta \leftarrow rint(0, L);$
$U \leftarrow \{i | 1 \le i \le M\};$
$C \leftarrow \{i | 1 \le i \le \eta\};$
$\forall 1 \le i \le \eta, W[i] \leftarrow 0;$
while $U \ne \emptyset \wedge C \ne \emptyset$ **do**
  $i \leftarrow U[rint(1, |U|)];$
  $q \leftarrow rand()\lambda_i;$
  $f \leftarrow q;$
  $C' \leftarrow C;$
  while $f > 0 \wedge C' \ne \emptyset$ **do**
    $j \leftarrow C'[rint(1, |C'|)];$
    $C' \leftarrow C' - \{j\};$
    $f' \leftarrow \min\{f, rand()q\};$
    $f' \leftarrow \min\{\mu - W[j], f'\beta_i\}/\beta_i;$
    $S.Z[i][j] \leftarrow S.Z[i][j] + f'/\lambda_i;$
    $f \leftarrow f - f';$
    $W[j] \leftarrow W[j] + f'\beta_i;$
    if $W[j] \ge \mu$ **then**
      $C \leftarrow C - \{j\};$
    end if
  end while
  if $C \ne \emptyset \wedge f > 0$ **then**
    $C' \leftarrow \{k | k \in C \wedge W[k] + f\beta_i \le \mu\};$
    if $C' \ne \emptyset$ **then**
      $j \leftarrow C'[rint(1, |C'|)];$
      $S.Z[i][j] \leftarrow S.Z[i][j] + f/\lambda_i;$
      $W[j] \leftarrow W[j] + f\beta_i;$
      if $W[j] \ge \mu$ **then**
        $C \leftarrow C - \{j\};$
      end if
    end if
  end if
end while
$\forall 1 \le i \le M, S.Y[i] \leftarrow (1 - \sum_{1 \le j \le \eta} S.Z[i][j])\lambda_i;$
$A \leftarrow \{i | 1 \le i \le N\};$
$C \leftarrow \{i | 1 \le i \le \eta \wedge W[i] > 0\};$
for $i \in C$ **do**
  $S.X[i] \leftarrow A[rint(1, |A|)];$
  $A \leftarrow A - \{S.X[i]\};$
end for
return S;

the constraint of their total offloaded task flows determined by their workloads. Then, they are further modified subject to the maximum workload constraint of each cloudlet. After all task offloading probabilities of all users are determined, their workloads need to be adjusted such that the sum of the workload and total offloaded task flow of each of them is equal to its task arrival rate. Finally, cloudlet deployment is modified such that each cloudlet without any workload is not deployed and no two cloudlets are deployed at the same site.

A solution repairing (SR) algorithm is illustrated in Algorithm 2, where lines 2 and 3 restrict the cloudlet deployment and user workload components within their bounds. The values of the flows of the tasks offloaded to the cloudlets that are not deployed are cleared to zero at line 4. The for-loop at line 5 revises the task offloading probabilities of users based on their workloads. Line 6 computes the total task offloading probability for a user, and line 7 restricts its probabilities. The while-loop at line 11 finds the offloaded task flows from a given user that should be reduced, and lines 15 and 16 reduce them. The violation of the maximum cloudlet workload is handled by the for-loop at line 19. The while-loop at line 24 finds the offloaded task flows to a given cloudlet that should be decreased, and lines 28 and 29 decrease them. The workloads of users are reset according to their final task offloading probabilities at line 33. Line 34 states that if no tasks are offloaded to a cloudlet, it does not need to be deployed. Lines 35–49 deal with the case that more than one cloudlet is placed at a site. Specially, the while-loop at line 36 determines the set of the cloudlets with valid deployment at random. The while-loop at line 44 places each cloudlet with invalid deployment at a randomly selected unused site.

In SR, it needs $O(ML)$ time to run lines 2–4. The while-loop at line 11 requires $O(L)$ time, and thus the for-loop at line 5 consumes $O(ML)$ time. The for-loop at line 19 consumes $O(ML)$ time, since the while-loop at line 24 takes $O(M)$ time. Lines 32–34 take $O(ML)$ time. It takes $O(L)$ time to execute the while-loop at line 36. The while-loop at line 44 consumes $O(L)$ time, because it has at most $O(L)$ iterations. Therefore, the time complexity of SR is $O(4ML + 2L)$=$O(ML)$.

4.5 Generalized and Quasi-opposition-Based Learning

In order to avoid trapping into local optima and accelerate convergence, differently from the application of the opposition-based learning (OBL) [31, 52] in [57], we make comprehensive use of the generalized opposition-based learning (GOBL) [55] and the quasi-opposition-based learning (QOBL) [43, 54] to produce the opposites of whale positions, both of which are an extension of OBL [31, 52]. By introducing GOBL and QOBL, not only whale positions but also their opposites are used to update the archive. Therefore, the diversity of nondominated solutions stored in the archive can be improved, and the searching behavior of whales can be better guided.

In OBL [31, 52], GOBL [55], and QOBL [43, 54], the opposite numbers of a component q_i of a vector $Q = (q_1, q_2, \ldots, q_n)^{\mathrm{T}}$ are respectively defined as follows:

Algorithm 2 SR

```
Input: whale position S
Output: whale position S′
S′ ← S;
∀1 ≤ i ≤ L, S′.X[i] ← max{min{S′.X[i], N}, 2 − N};
∀1 ≤ i ≤ M, S′.Y[i] ← max{min{S′.Y[i], λ_i}, 0};
∀1 ≤ i ≤ L ∧ S′.X[i] ≤ 0, ∀1 ≤ j ≤ M, S′.Z[j][i] ← 0;
for i ← 1 to M do
   ρ ← (λ_i − S′.Y[i])/λ_i;
   ∀1 ≤ j ≤ L, S′.Z[i][j] ← max{min{S′.Z[i][j], ρ}, 0};
   if Σ_{1≤j≤L∧S′.X[j]>0} S′.Z[i][j] > ρ then
      C ← {j|1 ≤ j ≤ L ∧ S′.X[j] > 0 ∧ S′.Z[i][j] > 0};
      q ← 0; h ← 0;
      while q < ρ do
         h ← C[rint(1, |C|)];
         q ← q + S′.Z[i][h]; C ← C − {h};
      end while
      S′.Z[i][h] ← S′.Z[i][h] − (q − ρ);
      ∀j ∈ C, S′.Z[i][j] ← 0;
   end if
end for
for i ∈ {i|1 ≤ i ≤ L ∧ S′.X[i] > 0} do
   f ← Σ_{1≤j≤M} S′.Z[j][i]λ_jβ_j;
   if f > μ then
      U ← {j|1 ≤ j ≤ M ∧ S′.Z[j][i] > 0};
      q ← 0; h ← 0;
      while q < μ do
         h ← U[rint(1, |U|)];
         q ← q + S′.Z[h][i]λ_hβ_h; U ← U − {h};
      end while
      S′.Z[h][i] ← S′.Z[h][i] − (q − μ)/(β_hλ_h);
      ∀j ∈ U, S′.Z[j][i] ← 0;
   end if
end for
∀1 ≤ i ≤ L, W[i] ← Σ_{1≤j≤M} S′.Z[j][i]λ_jβ_j;
∀1 ≤ i ≤ M, S′.Y[i] ← (1 − Σ_{1≤j≤L} S′.Z[i][j])λ_i;
∀1 ≤ i ≤ L ∧ S′.X[i] > 0 ∧ W[i] ≤ 0, S′.X[i] ← 0;
A ← ∅; C ← ∅; C′ ← {i|1 ≤ i ≤ L ∧ S′.X[i] > 0};
while C′ ≠ ∅ do
   i ← C′[rint(1, |C′|)]; C′ ← C′ − {i};
   if S′.X[i] ∉ A then
      A ← A ∪ {S′.X[i]}; C ← C ∪ {i};
   end if
end while
A′ ← {i|1 ≤ i ≤ N ∧ i ∉ A};
C″ ← {i|1 ≤ i ≤ L ∧ S′.X[i] > 0 ∧ i ∉ C};
while C″ ≠ ∅ do
   i ← C″[rint(1, |C″|)];
   C″ ← C″ − {i};
   S′.X[i] ← A′[rint(1, |A′|)];
   A′ ← A′ − {S′.X[i]};
end while
return S′;
```

$$\bar{q}_i^o = \check{q}_i + \hat{q}_i - q_i \tag{23}$$

$$\bar{q}_i^{go} = rand()(\check{q}_i + \hat{q}_i) - q_i \tag{24}$$

$$\bar{q}_i^{qo} = (\check{q}_i + \hat{q}_i)/2 + rand()(\bar{q}_i^o - (\check{q}_i + \hat{q}_i)/2) \tag{25}$$

where $\check{q}_i$ and $\hat{q}_i$ are the lower and upper bounds of q_i, respectively.

By combining QOBL and GOBL, we can obtain another opposite number of a component q_i, which is defined as

$$\bar{q}_i^{co} = (\check{q}_i + \hat{q}_i)/2 + rand()(\bar{q}_i^{go} - (\check{q}_i + \hat{q}_i)/2). \tag{26}$$

In the opposite $\bar{S}_i$ of a whale position S_i, the opposite number $\bar{S}_i.X[j]$ of $S_i.X[j]$ is calculated by GOBL, while the opposite numbers $\bar{S}_i.Y[j]$ and $\bar{S}_i.Z[j][k]$ of $S_i.Y[j]$ and $S_i.Z[j][k]$ are computed by the combination of QOBL and GOBL. Therefore, $\bar{S}_i.X[j] = 2rand() - S_i.X[j]$, $\bar{S}_i.Y[j] = \lambda_j/2 + rand()(rand()\lambda_j - S_i.Y[j] - \lambda_j/2)$, and $\bar{S}_i.Z[j][k] = 1/2 + rand()(rand() - S_i.Z[j][k] - 1/2)$.

4.6 *MGW for MOCT*

MGW utilizes a number of whales to search for the true Pareto front. After the whale positions are randomly initialized by RFSCA, their opposites are computed and repaired. Both the initial positions of whales and their opposites are employed to initialize the archive.

In order to improve the exploration and exploitation abilities of whales, the differential mutation and crossover in differential evolution (DE) [20, 49, 50, 65] are employed to modify the whale foraging behavior. If the DE operations are selected to update a whale position, in order to balance exploration and exploitation, three mutually different solutions $\mathbb{S}$, $\mathbb{S}'$, and $\mathbb{S}''$ or two ones $\mathbb{S}$ and $\mathbb{S}'$ need to be randomly selected from the archive before the updating. The probabilities of selecting the three and two ones are equal. The crossover [49] is performed on the mutant result and $\mathbb{S}$. A random component of the whale position is updated using the differential mutation definitely. For each of the other components k of the whale position, it is updated using the differential mutation with the probability of CR and equals $\mathbb{S}[k]$ with the probability of $1-CR$. Denote F as the scaling factor. When the differential mutation is chosen, if $\mathbb{S}$, $\mathbb{S}'$, and $\mathbb{S}''$ are selected, a whale position component $S_i[j]$ is updated as [49]

$$S_i^{t+1}[j] = \mathbb{S}[j] + F(\mathbb{S}'[j] - \mathbb{S}''[j]); \tag{27}$$

otherwise, it is updated as

$$S_i^{t+1}[j] = \mathbb{S}[j] + F(\mathbb{S}'[j] - \mathbb{S}[j]). \tag{28}$$

During the search process, each whale position is updated with the prey encircling, spiral movement, prey searching, or DE operations. If the first two methods are adopted, a leader S^* has to be randomly selected from a number of the solutions with the biggest crowding distances in the archive first, whose proportion is denoted by ς. A random whale ought to be first selected from the current population when the prey searching is used. If the DE operations are chosen, three or two mutually different solutions should be first selected from the archive at random. If a new position is invalid, it is repaired by SR. After the positions of all whales are updated, they are used to update the archive. In each generation, the opposites of the whale positions are computed with the probability of ϱ, which are also used to update the archive. The new archive contains the nondominated solutions identified from the whale positions, their opposites, and the solutions in the old archive. If the size of the new archive exceeds its capacity, some solutions with the smallest crowding distances are removed. Denote Υ and Γ as the archive capacity and the number of generations, respectively. In each generation t, $a = 2(1 - ((t-1)/\Gamma)^3)$. MGW is summarized in Algorithm 3, and its flowchart is shown in Fig. 2.

In MGW, line 1 consumes $O(MLH)$ time. Both for-loops at lines 2 and 11 take $O(MLH)$ time. Both lines 9 and 10 require $O(\Upsilon log_2\Upsilon)$ time. Because the crowding distance assignment requires $O((H + \Upsilon)log_2(H + \Upsilon))$ time [11] and it takes $O(H(H + \Upsilon))$ time to identify the nondominated solutions from the archive and the whale population, the archive updating at lines 42 and 44 consumes $O((H+\Upsilon)(H+log_2(H+\Upsilon)))$ time. As a result, it takes $O(\Gamma(\Upsilon log_2\Upsilon + MLH + (H + \Upsilon)(H + log_2(H + \Upsilon)))) = O(\Gamma(MLH + (H + \Upsilon)(H + log_2(H + \Upsilon))))$ time to run the for-loop at line 7. Therefore, the time complexity of MGW is $O(\Gamma(MLH + (H + \Upsilon)(H + log_2(H + \Upsilon))))$.

5 Simulation Results

5.1 Comparison and Metrics

In this section, we compare our proposed MGW with NSGA-II [11], GPAWOA [21], multi-objective differential evolution (MODE) [48], multi-objective grey wolf optimizer (MOGWO) [40], and multi-objective particle swarm optimization (MOPSO) [6] adapted to handle MOCT. In them, the approaches of encoding, initializing, and repairing individuals are the same as those in MGW. NSGA-II [11], MOPSO [6], and MODE [48] are classic and commonly used methods, while GPAWOA [21] and MOGWO [40] are very recent ones. The comparison is in terms of the inverted generational distance (IGD) [19, 36, 66] and hypervolume (HV) [5, 47], which are the two most effective performance indicators to evaluate the

Fig. 2 Flowchart of MGW

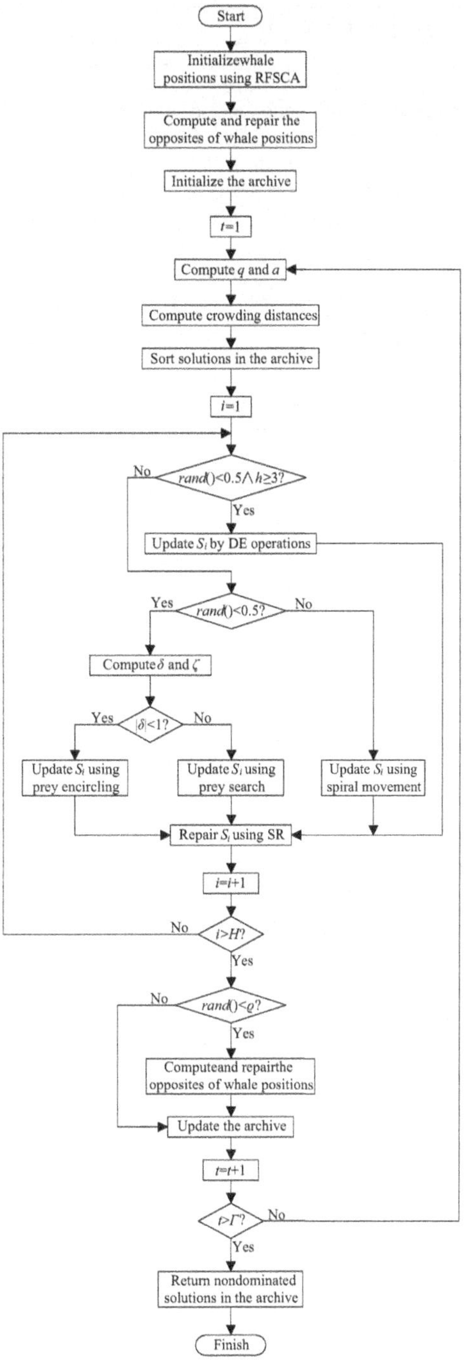

Algorithm 3 MGW

Input: $u_1, u_2, \ldots, u_M; v_1, v_2, \ldots, v_N; c_1, c_2, \ldots, c_L; \vartheta_1, \vartheta_2, \ldots, \vartheta_M; \lambda_1, \lambda_2, \ldots, \lambda_M; \alpha_1, \alpha_2, \ldots, \alpha_M; \beta_1, \beta_2, \ldots, \beta_M; \varphi_1^u, \varphi_2^u, \ldots, \varphi_M^u; B; \varphi^c; \mu$
Output: Set of nondominated solutions

$\forall 1 \leq i \leq H, S_i \leftarrow RFSCA()$;
for $i \leftarrow 1$ to H **do**
    Compute the opposite $\bar{S}_i$ of S_i;
    $\bar{S}_i \leftarrow SR(\bar{S}_i)$;
end for
Initialize the archive using $\{S_i | 1 \leq i \leq H\} \cup \{\bar{S}_i | 1 \leq i \leq H\}$;
for $t \leftarrow 1$ to Γ **do**
    $q \leftarrow \lceil \varsigma h \rceil$, where h is the archive size; $a \leftarrow 2(1 - ((t-1)/\Gamma)^3)$;
    Compute the crowding distances of the solutions in the archive;
    Sort the solutions in the archive in descending order by their crowding distances;
    for $i \leftarrow 1$ to H **do**
        if $rand() < 0.5$ and $h \geq 3$ **then**
            if $rand() < 0.5$ **then**
                Select three mutually different solutions $\mathbb{S}$, $\mathbb{S}'$, and $\mathbb{S}''$ from the archive randomly;
                Update S_i using $\mathbb{S}$, $\mathbb{S}'$, $\mathbb{S}''$, (27), and crossover;
            else
                Select two different solutions $\mathbb{S}$ and $\mathbb{S}'$ from the archive randomly;
                Update S_i using $\mathbb{S}$, $\mathbb{S}'$, (28), and crossover;
            end if
        else
            if $rand() < 0.5$ **then**
                Compute δ and ζ;
                if $|\delta| < 1$ **then**
                    Select a leader S^* from the first q solutions in the sorted archive at random;
                    Update S_i using S^* and (20);
                else
                    Choose a whale j randomly;
                    Update S_i using S_j and (22);
                end if
            else
                Select a leader S^* from the q solutions with the biggest crowding distances in the archive at random;
                Update S_i using S^* and (21);
            end if
        end if
        $S_i \leftarrow SR(S_i)$;
    end for
    if $rand() < \varrho$ **then**
        for $i \leftarrow 1$ to H **do**
            Compute the opposite $\bar{S}_i$ of S_i;
            $\bar{S}_i \leftarrow SR(\bar{S}_i)$;
        end for
        Update the archive using $\{S_i | 1 \leq i \leq H\} \cup \{\bar{S}_i | 1 \leq i \leq H\}$;
    else
        Update the archive using $\{S_i | 1 \leq i \leq H\}$;
    end if
end for
return nondominated solutions in the archive;

accuracy of a nondominated solution set. Lower IGD or larger HV of S indicates that it has higher quality.

The IGD of a nondominated solution set S is defined as the average distance between each point in the true Pareto front $\urcorner$ and its nearest point in the Pareto front of S, i.e. [19, 36, 66],

$$IGD(S) = \frac{1}{|\urcorner|} \sum_{(h_1,h_2,\ldots,h_m)\in\urcorner} \min_{Q\in S}\{ \\ (\sum_{1\le i\le m} (h_i - f_i(Q))^2)^{\frac{1}{2}}\} \tag{29}$$

The HV of S measures how much objective space S covers, i.e. [5, 47],

$$HV(S) = \beth(\cup_{Q\in S}\{(h_1, h_2, \ldots, h_m)|(\forall 1 \le i \le m, \\ f_i(Q) \le h_i \le h_i')\}) \tag{30}$$

where $\beth(\cdot)$ is the Lebesgue measure and $(h_1', h_2', \ldots, h_m')$ is the reference point.

Since the true Pareto front is unknown, the Pareto front extracted from the results of all the six algorithms is used as the substitute of the true Pareto front to compute IGD. Let $\check{\psi}_i$ and $\hat{\psi}_i$ $(1 \le i \le 2)$ denote the minimum and maximum values of the ith objective of the results of all the six algorithms, respectively. To compute IGD and HV, the objective values $\psi_1(S)$, $\psi_2(S)$, and $\psi_3(S)$ of each solution S are normalized as $\psi_1'(S) = (\hat{\psi}_1 - \psi_1(S))/(\hat{\psi}_1 - \check{\psi}_1)$, $\psi_2'(S) = (\hat{\psi}_2 - \psi_2(S))/(\hat{\psi}_2 - \check{\psi}_2)$, and $\psi_3'(S) = (L - \psi_3(S))/L$, respectively. For computing HV, the reference point is chosen to be $(-1/(\iota - 1), -1/(\iota - 1), -1/(\iota - 1))$, where ι is the number of all different network deployment costs in the results of all the six algorithms, according to the recommendation in [5]. We use Java to develop the simulation program on MyEclipse 2014 and run it on a Lenovo ThinkPad T450s, which has 2.20 GHz CPU and 4 GB physical RAM.

5.2 Parameter Settings

All APs and users are located in a 10,000 $\times$ 10,000 m area, which is divided into $N \times N$ grids. There is an AP in the center of each grid. Each user is randomly connected to an AP, and the connectivity graph of all APs is randomly generated. The position of each user is randomly generated, and the distance between each user and its associated AP is no more than $5000/N$m. The average input data size and computational resource demand per bit data of a task of each user are two random values in the ranges of [200 $\times$ 1024, 500 $\times$ 1024] bytes and [50, 100] cycles/bit [10], respectively. The average task arrival rate of each user is randomly drawn from [0.1, 3] tasks/second. The effective switched capacitance of CPU, computing capacity, and transmission power of each user are set to 5×10^{-27} [53], 2 GHz, and 0.1 W, respectively. The system bandwidth is set to 40 MHz, and the channel gain from a user to its associated AP is set to d^{-4}, where d is the distance between them. The maximum workload μ of each cloudlet is set to $0.9\varphi^c$. The background noise power is set to -100 dBm, and the link transmission rate between two APs is randomly generated in $[10^8, 2 \times 10^8]$ bits/second.

In NSGA-II, the simulated binary crossover and polynomial mutation are used. The crossover and mutation probabilities and the distribution indexes for crossover and mutation are set to 0.9, $1/(L + M(L + 1))$, 20, and 20 following [11], respectively. In MOPSO, the mutation rate is set to 0.5 following [6]. In both MOPSO and MOGWO, the number of the subdivisions for each dimension in the grid mechanism is set to 30 [6]. The parameters' setting of GPAWOA is the same as that in [21]. In MGW, b, ς, ϖ, and ϱ are set to 3, $3/\Upsilon$, 3, and 0.15, respectively. In both MODE and MGW, the scaling factor and crossover probability are set to 0.5 and 0.9, respectively. In MOPSO, MOGWO, GPAWOA, and MGW, the archive capacity is set to 100. In all algorithms, the population size and the number of generations are set to 100 and 2000, respectively.

5.3 *Results and Discussions*

In the simulations, we consider different M, L, N, and φ^c. For each setting of the parameters, 30 instances are randomly generated. IGDs and HVs of the results of each algorithm for the 30 instances are averaged. Tables 2, 3, 4, and 5 display the simulation results. From them, we can see that the performance of MGW with respect to IGD and HV is better than that of other algorithms. This means that MGW can produce a better nondominated solution set than the others. In other words, the Pareto front obtained by MGW is closest to the true Pareto front. Meanwhile, the running time of MGW is only longer than that of GPAWOA and MOGWO. From Tables 2 and 3, we can observe that the running time of the algorithms increases with the growth of the upper bound of the budget or the number of users. The reason is that larger upper bound of the budget or more users result in a bigger problem size. As shown in Table 5, with the increase of the cloudlet computing capacity, IGD of the result of MGW decreases and its HV increases. This is due to the fact that larger cloudlet computing capacity permits shorter task response delay and less energy consumption of users and fewer deployed cloudlets and thus makes the normalized objective values of solutions increase.

The Pareto fronts obtained by the algorithms for two random instances are shown in Figs. 3 and 4, respectively. For each instance, each algorithm is run 30 times independently, and the 100 best nondominated solutions are selected from the results of the 30 runs using the crowded-comparison operator in [11]. In Fig. 3, the IGDs of the Pareto fronts delivered by MOPSO, MOGWO, NSGA-II, MODE, GPAWOA, and MGW are 0.1457, 0.2024, 0.1978, 0.0265, 0.1875, and 0.0211, respectively,

Table 2 Performance comparison with $M = 240$, $N = 120$, and $\varphi^c = 25$ GHz under different upper bound of the budget

L	Method	IGD	HV	Time (s)
30	NSGA-II	0.2329	0.5762	235.65
	MOPSO	0.2098	0.6384	158.13
	MOGWO	0.2432	0.5565	64.9
	MODE	0.0454	0.9381	109.2
	GPAWOA	0.2281	0.5718	49.12
	MGW	0.0267	0.9423	98.55
35	NSGA-II	0.2393	0.5696	286.95
	MOPSO	0.223	0.6059	203.46
	MOGWO	0.2575	0.5366	88.27
	MODE	0.0446	0.9294	138.02
	GPAWOA	0.2335	0.5671	58.58
	MGW	0.0241	0.9405	122.25
40	NSGA-II	0.2321	0.5618	312.49
	MOPSO	0.2201	0.6023	220.6
	MOGWO	0.2483	0.5366	89.04
	MODE	0.04	0.9252	142.95
	GPAWOA	0.2243	0.5608	60.38
	MGW	0.0253	0.9401	129.26
45	NSGA-II	0.2475	0.5585	340.05
	MOPSO	0.2508	0.5731	242.51
	MOGWO	0.2628	0.5294	93.91
	MODE	0.048	0.9263	150.87
	GPAWOA	0.2314	0.5602	65.09
	MGW	0.029	0.9326	138.91
50	NSGA-II	0.2426	0.5581	377.98
	MOPSO	0.2504	0.5702	273.89
	MOGWO	0.2593	0.5244	103.72
	MODE	0.0487	0.9248	166.29
	GPAWOA	0.2379	0.548	73.25
	MGW	0.0311	0.9287	154.35

and their HVs are 0.7216, 0.5712, 0.6033, 0.9336, 0.5842, and 0.9444, respectively. In Fig. 4, the IGDs of the Pareto fronts obtained by MOPSO, MOGWO, NSGA-II, MODE, GPAWOA, and MGW are 0.2444, 0.2356, 0.2242, 0.0292, 0.2216, and 0.0104, respectively, and their HVs are 0.562, 0.5428, 0.5689, 0.9359, 0.5561, and 0.9472, respectively. From the figures, we can observe that the Pareto front obtained by MGW has better accuracy and diversity than those obtained by other algorithms, because it is distributed over a larger area and covers more objective space. This is because the effective combination of GPAWOA, GOBL, QOBL, and DE enables MGW to have stronger exploration and exploitation abilities. From the figures, it can be observed that by and large, the average task response delay of users and the network deployment cost grow with the reduction of the average energy

Table 3 Performance comparison with $L = 40$, $N = 120$, and $\varphi^c = 25\,\text{GHz}$ under different number of users

M	Method	IGD	HV	Time (s)
200	NSGA-II	0.2637	0.569	250.9
	MOPSO	0.2443	0.6098	173.53
	MOGWO	0.282	0.538	66.5
	MODE	0.0435	0.9629	105.8
	GPAWOA	0.2586	0.558	41.61
	MGW	0.0211	0.9703	97.18
220	NSGA-II	0.252	0.5699	286.94
	MOPSO	0.2489	0.5858	200.39
	MOGWO	0.2729	0.5393	79.2
	MODE	0.0453	0.947	123.29
	GPAWOA	0.2454	0.5598	50.95
	MGW	0.0254	0.9519	112.67
240	NSGA-II	0.2295	0.5609	313.5
	MOPSO	0.2167	0.6014	218.94
	MOGWO	0.2493	0.5353	87.64
	MODE	0.0453	0.9152	137.64
	GPAWOA	0.2177	0.5602	54.94
	MGW	0.0257	0.9331	123.8
260	NSGA-II	0.2273	0.563	328.06
	MOPSO	0.2274	0.5899	225.77
	MOGWO	0.2456	0.5323	90.57
	MODE	0.0526	0.9073	140.81
	GPAWOA	0.2191	0.5579	58.11
	MGW	0.0317	0.922	129.77
280	NSGA-II	0.2181	0.5499	357.14
	MOPSO	0.2219	0.5615	248.3
	MOGWO	0.2295	0.5295	98.0
	MODE	0.0542	0.884	156.22
	GPAWOA	0.2078	0.5498	63.93
	MGW	0.0358	0.8905	144.3

consumption of users, and the average task response delay of users declines with the increase of the number of deployed cloudlets. This observation verifies that there exist conflicts among the objectives of MOCT. As shown in the figures, MGW can deliver solutions with lower-energy consumption and task response delay of users and number of deployed cloudlets. This means that MGW can better optimize user QoE and network deployment cost, which can also be verified by the higher accuracy of MGW revealed by the tables.

Table 4 Performance comparison with $M = 240$, $L = 40$, and $\varphi^c = 25$ GHz under different number of APs

N	Method	IGD	HV	Time (s)
100	NSGA-II	0.2412	0.5601	286.13
	MOPSO	0.2495	0.5652	213.9
	MOGWO	0.2574	0.5363	80.86
	MODE	0.042	0.9253	131.5
	GPAWOA	0.2363	0.5577	52.29
	MGW	0.0289	0.9279	117.59
110	NSGA-II	0.2351	0.5619	287.29
	MOPSO	0.2338	0.585	210.61
	MOGWO	0.2512	0.5393	81.38
	MODE	0.0469	0.9212	132.92
	GPAWOA	0.229	0.5549	51.52
	MGW	0.0275	0.9349	118.1
120	NSGA-II	0.2427	0.562	286.53
	MOPSO	0.2366	0.6018	207.81
	MOGWO	0.26	0.535	79.81
	MODE	0.044	0.9318	131.04
	GPAWOA	0.2385	0.5566	50.17
	MGW	0.028	0.9387	115.69
130	NSGA-II	0.2401	0.5665	285.42
	MOPSO	0.2405	0.5886	210.01
	MOGWO	0.251	0.5412	80.27
	MODE	0.0452	0.9313	132.96
	GPAWOA	0.2319	0.559	51.51
	MGW	0.0265	0.9386	118.77
140	NSGA-II	0.2387	0.5634	291.43
	MOPSO	0.2335	0.5898	213.01
	MOGWO	0.2547	0.5335	83.34
	MODE	0.0426	0.9313	136.59
	GPAWOA	0.2297	0.5561	53.33
	MGW	0.0245	0.941	119.7

6 Summary

In this work, the problem of joint cloudlet deployment and task offloading is addressed. Its objectives are to minimize the energy consumption and task response delay of users and the number of deployed cloudlets. This optimization problem is intractable and formulated as a mixed integer nonlinear program. To deal with this NP-complete problem, a modified guided population archive whale optimization algorithm is presented, and its time complexity is analyzed as well.

Table 5 Performance comparison with $M = 240$, $L = 40$, and $N = 120$ under different cloudlet computing capacity (GHz)

φ^c (GHz)	Method	IGD	HV	Time (s)
15	NSGA-II	0.1796	0.543	292.93
	MOPSO	0.2191	0.5047	216.99
	MOGWO	0.2045	0.5102	88.04
	MODE	0.0578	0.8192	143.87
	GPAWOA	0.1712	0.5426	56.05
	MGW	0.0466	0.8266	126.59
20	NSGA-II	0.2071	0.5558	292.94
	MOPSO	0.2184	0.5729	217.78
	MOGWO	0.2321	0.5276	85.19
	MODE	0.0539	0.8697	140.69
	GPAWOA	0.2038	0.5561	55.04
	MGW	0.0422	0.8736	124.55
25	NSGA-II	0.2376	0.5555	291.9
	MOPSO	0.2367	0.5871	214.85
	MOGWO	0.2606	0.5238	83.97
	MODE	0.0489	0.9187	135.18
	GPAWOA	0.2322	0.5535	53.79
	MGW	0.026	0.9328	118.35
30	NSGA-II	0.2579	0.568	303.19
	MOPSO	0.248	0.6002	227.54
	MOGWO	0.2827	0.5319	92.0
	MODE	0.0378	0.9643	146.02
	GPAWOA	0.2541	0.557	56.98
	MGW	0.0177	0.9759	125.62
35	NSGA-II	0.2521	0.5612	294.94
	MOPSO	0.2279	0.6148	216.43
	MOGWO	0.2762	0.5251	84.68
	MODE	0.0359	0.9838	139.46
	GPAWOA	0.2492	0.541	53.08
	MGW	0.0155	0.9965	120.11

In this algorithm, the effective methods of encoding, initializing, and repairing a whale position are designed. Besides, the generalized and quasi-opposition-based learning and the mutation and crossover in differential evolution are adopted to improve its optimization ability. Compared with other meta-heuristic algorithms, the proposed algorithm can deliver a nondominated solution set with higher quality with competitive efficiency. Our future work includes designing a scheme to optimize channel bandwidth allocation to users and testing other multi-objective optimization methods based on recently developed algorithms, e.g., [8, 9, 23, 24, 27, 51, 67, 68].

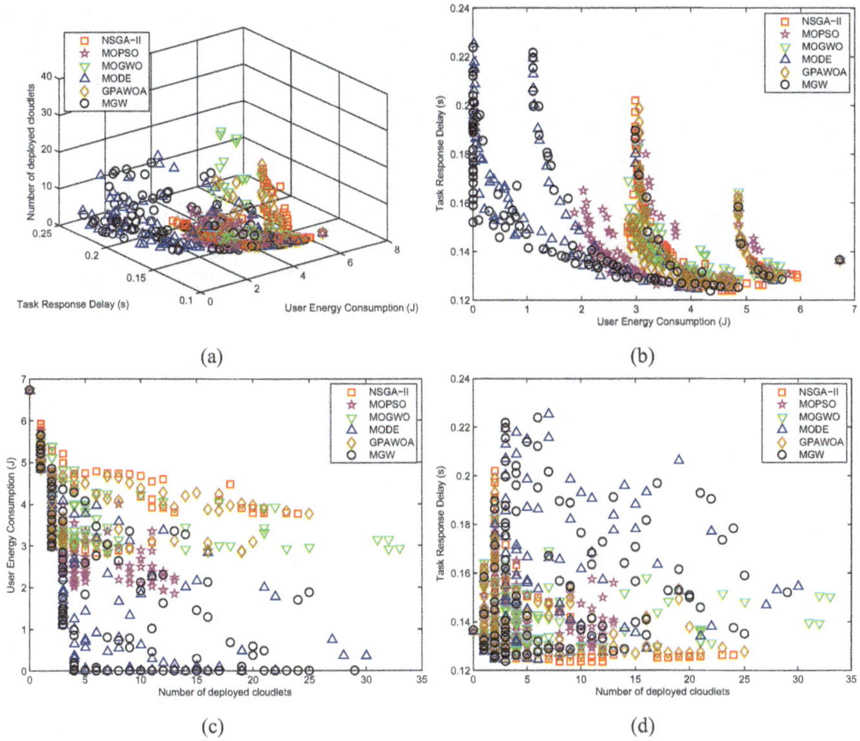

Fig. 3 Pareto fronts obtained by different algorithms for a random instance with $M = 240$, $N = 120$, $L = 40$, and $\varphi^c = 25$ GHz. (**a**) Visualization with three objectives. (**b**) Visualization with ψ_1 and ψ_2. (**c**) Visualization with ψ_1 and ψ_3. (**d**) Visualization with ψ_2 and ψ_3

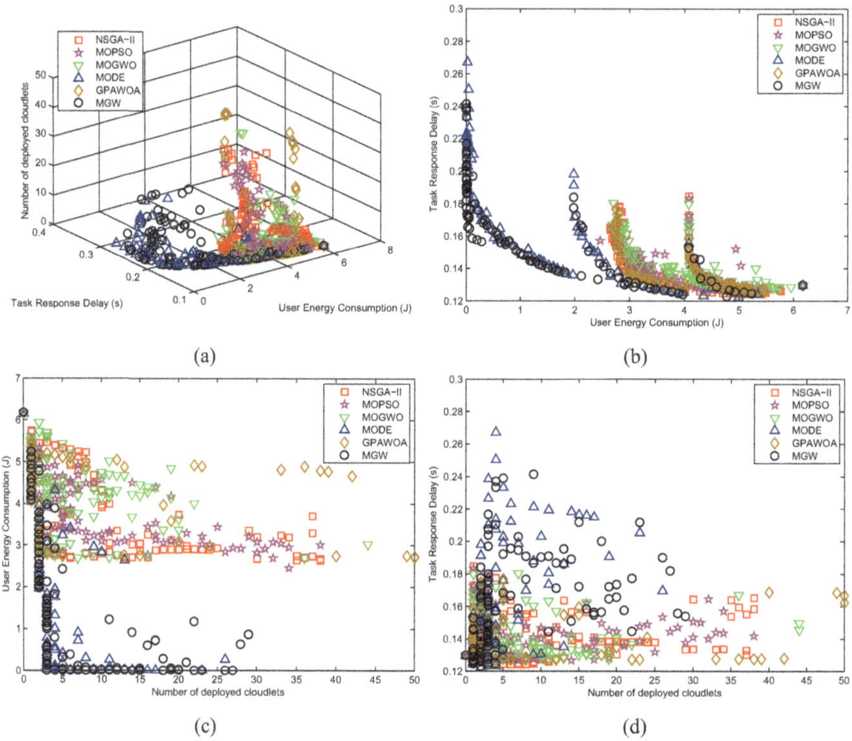

Fig. 4 Pareto fronts obtained by different algorithms for a random instance with $M = 300$, $N = 150$, $L = 50$, and $\varphi^c = 35\,\text{GHz}$. (**a**) Visualization with three objectives. (**b**) Visualization with ψ_1 and ψ_2. (**c**) Visualization with ψ_1 and ψ_3. (**d**) Visualization with ψ_2 and ψ_3

References

1. N. Abbas, Y. Zhang, A. Taherkordi, T. Skeie, Mobile edge computing: a survey. IEEE Internet Things J. **5**(1), 450–465 (2018)
2. G. Aloi, G. Fortino, R. Gravina, P. Pace, C. Savaglio, Simulation-driven platform for edge-based AAL systems. IEEE J. Sel. Areas Commun. **39**(2), 446–462 (2021)
3. A. Barbieri, F. Marozzo, C. Savaglio, IoT platforms and services configuration through parameter sweep: a simulation-based approach, in *2021 IEEE International Conference on Systems, Man, and Cybernetics (SMC)* (2021), pp. 1803–1808
4. J. Bi, H. Yuan, S. Duanmu, M. Zhou, A. Abusorrah, Energy-optimized partial computation offloading in mobile-edge computing with genetic simulated-annealing-based particle swarm optimization. IEEE Internet Things J. **8**(5), 3774–3785 (2021)
5. Y. Cao, B.J. Smucker, T.J. Robinson, On using the hypervolume indicator to compare Pareto fronts: applications to multi-criteria optimal experimental design. J. Statist. Plann. Inference **160**, 60–74 (2015)
6. C.A.C. Coello, G.T. Pulido, M.S. Lechuga, Handling multiple objectives with particle swarm optimization. IEEE Trans. Evol. Comput. **8**(3), 256–279 (2004)

7. L. Cui, C. Xu, S. Yang, J.Z. Huang, J. Li, X. Wang, Z. Ming, N. Lu, Joint optimization of energy consumption and latency in mobile edge computing for internet of things. IEEE Internet Things J. **6**(3), 4791–4803 (2019)
8. M. Cui, L. Li, M. Zhou, A. Abusorrah, Surrogate-assisted autoencoder-embedded evolutionary optimization algorithm to solve high-dimensional expensive problems. IEEE Trans. Evol. Comput. **26**(4), 676–689 (2022)
9. M. Cui, L. Li, M. Zhou, J. Li, A. Abusorrah, K. Sedraoui, A bi-population cooperative optimization algorithm assisted by an autoencoder for medium-scale expensive problems. IEEE/CAA J. Automat. Sin. **9**(11), 1952–1966 (2022)
10. Y. Dai, D. Xu, S. Maharjan, Y. Zhang, Joint computation offloading and user association in multi-task mobile edge computing. IEEE Trans. Veh. Technol. **67**(12), 12313–12325 (2018)
11. K. Deb, A. Pratap, S. Agarwal, T. Meyarivan, A fast and elitist multiobjective genetic algorithm: NSGA-II. IEEE Trans. Evol. Comput. **6**(2), 182–197 (2002)
12. E. El Haber, T.M. Nguyen, C. Assi, Joint optimization of computational cost and devices energy for task offloading in multi-tier edge-clouds. IEEE Trans. Commun. **67**(5), 3407–3421 (2019)
13. Q. Fan, N. Ansari, Cost aware cloudlet placement for big data processing at the edge, in *2017 IEEE International Conference on Communications (ICC)* (2017), pp. 1–6
14. Q. Fan, N. Ansari, Application aware workload allocation for edge computing-based IoT. IEEE Internet Things J. **5**(3), 2146–2153 (2018)
15. Q. Fan, N. Ansari, Workload allocation in hierarchical cloudlet networks. IEEE Commun. Lett. **22**(4), 820–823 (2018)
16. Q. Fan, N. Ansari, On cost aware cloudlet placement for mobile edge computing. IEEE/CAA J. Automat. Sin. **6**(4), 926–937 (2019)
17. Q. Fan, N. Ansari, Towards workload balancing in fog computing empowered IoT. IEEE Trans. Netw. Sci. Eng. **7**(1), 253–262 (2020)
18. Y. Feng, M. Zhou, G. Tian, Z. Li, Z. Zhang, Q. Zhang, J. Tan, Target disassembly sequencing and scheme evaluation for CNC machine tools using improved multiobjective ant colony algorithm and fuzzy integral. IEEE Trans. Syst. Man Cybernet.: Syst. **49**(12), 2438–2451 (2019)
19. K. Gao, F. Yang, M. Zhou, Q. Pan, P.N. Suganthan, Flexible job-shop rescheduling for new job insertion by using discrete Jaya algorithm. IEEE Trans. Cybernet. **49**(5), 1944–1955 (2019)
20. S. Gao, Y. Yu, Y. Wang, J. Wang, J. Cheng, M. Zhou, Chaotic local search-based differential evolution algorithms for optimization. IEEE Trans. Syst. Man Cybernet. Syst. **51**(6), 3954–3967 (2021)
21. A. Got, A. Moussaoui, D. Zouache, A guided population archive whale optimization algorithm for solving multiobjective optimization problems. Expert Syst. Appl. **141**, 112972 (2020)
22. H. Guo, J. Zhang, J. Liu, H. Zhang, Energy-aware computation offloading and transmit power allocation in ultradense IoT networks. IEEE Internet Things J. **6**(3), 4317–4329 (2019)
23. X. Guo, M. Zhou, S. Liu, L. Qi, Lexicographic multiobjective scatter search for the optimization of sequence-dependent selective disassembly subject to multiresource constraints. IEEE Trans. Cybernet. **50**(7), 3307–3317 (2020)
24. S. Han, K. Zhu, M. Zhou, X. Liu, H. Liu, Y. Al-Turki, A. Abusorrah, A novel multiobjective fireworks algorithm and its applications to imbalanced distance minimization problems. IEEE/CAA J. Automat. Sin. **9**(8), 1476–1489 (2022)
25. N. Hassan, S. Gillani, E. Ahmed, I. Yaqoob, M. Imran, The role of edge computing in internet of things. IEEE Commun. Mag. **56**(11), 110–115 (2018)
26. Q. Hu, Y. Cai, G. Yu, Z. Qin, M. Zhao, G.Y. Li, Joint offloading and trajectory design for UAV-enabled mobile edge computing systems. IEEE Internet Things J. **6**(2), 1879–1892 (2019)
27. L. Huang, M. Zhou, K. Hao, Non-dominated immune-endocrine short feedback algorithm for multi-robot maritime patrolling. IEEE Trans. Intell. Transp. Syst. **21**(1), 362–373 (2020)
28. K. Intharawijitr, K. Iida, H. Koga, K. Yamaoka, Practical enhancement and evaluation of a low-latency network model using mobile edge computing, in *2017 IEEE 41st Annual Computer Software and Applications Conference (COMPSAC)*, vol. 1 (2017), pp. 567–574

29. S. Jeong, O. Simeone, J. Kang, Mobile edge computing via a UAV-mounted cloudlet: optimization of bit allocation and path planning. IEEE Trans. Veh. Technol. **67**(3), 2049–2063 (2018)
30. M. Jia, J. Cao, W. Liang, Optimal cloudlet placement and user to cloudlet allocation in wireless metropolitan area networks. IEEE Trans. Cloud Comput. **5**(4), 725–737 (2017)
31. Q. Kang, C. Xiong, M. Zhou, L. Meng, Opposition-based hybrid strategy for particle swarm optimization in noisy environments. IEEE Access **6**, 21888–21900 (2018)
32. A. Khalili, S. Zarandi, M. Rasti, Joint resource allocation and offloading decision in mobile edge computing. IEEE Commun. Lett. **23**(4), 684–687 (2019)
33. W.Z. Khan, E. Ahmed, S. Hakak, I.Yaqoob, A. Ahmed, Edge computing: a survey. Futur. Gener. Comput. Syst. **97**, 219–235 (2019)
34. K. Kumar, J. Liu, Y.H. Lu, B. Bhargava, A survey of computation offloading for mobile systems. Mob. Netw. Appl. **18**(1), 129–140 (2013)
35. L. Liu, Z. Chang, X. Guo, S. Mao, T. Ristaniemi, Multiobjective optimization for computation offloading in fog computing. IEEE Internet Things J. **5**(1), 283–294 (2018)
36. Z. Lv, L. Wang, Z. Han, J. Zhao, W. Wang, Surrogate-assisted particle swarm optimization algorithm with Pareto active learning for expensive multi-objective optimization. IEEE/CAA J. Automat. Sin. **6**(3), 838–849 (2019)
37. P. Mach, Z. Becvar, Mobile edge computing: a survey on architecture and computation offloading. IEEE Commun. Surv. Tutorials **19**(3), 1628–1656 (2017)
38. T. Mekonnen, M. Komu, R. Morabito, T. Kauppinen, E. Harjula, T. Koskela, M. Ylianttila, Energy consumption analysis of edge orchestrated virtualized wireless multimedia sensor networks. IEEE Access **6**, 5090–5100 (2018)
39. S. Mirjalili, A. Lewis, The whale optimization algorithm. Adv. Eng. Softw. **95**, 51–67 (2016)
40. S. Mirjalili, S. Saremi, S.M. Mirjalili, L. dos S. Coelho, Multi-objective grey wolf optimizer: A novel algorithm for multi-criterion optimization. Expert Syst. Appl. **47**, 106–119 (2016)
41. Pham, Q., Leanh, T., Tran, N.H., Park, B.J., Hong, C.S.: Decentralized computation offloading and resource allocation for mobile-edge computing: A matching game approach. IEEE Access **6**, 75868–75885 (2018)
42. G. Premsankar, B. Ghaddar, M. Di Francesco, R. Verago, Efficient placement of edge computing devices for vehicular applications in smart cities, in *2018 IEEE/IFIP Network Operations and Management Symposium (NOMS 2018)* (2018), pp. 1–9
43. S. Rahnamayan, H.R. Tizhoosh, M.M.A. Salama, Quasi-oppositional differential evolution, in *2007 IEEE Congress on Evolutionary Computation* (2007), pp. 2229–2236
44. Y. Ren, F. Zeng, W. Li, L. Meng, A low-cost edge server placement strategy in wireless metropolitan area networks, in *2018 27th International Conference on Computer Communication and Networks (ICCCN)* (2018), pp. 1–6
45. Z. Ren, H. Jiang, J. Xuan, Z. Luo, An accelerated-limit-crossing-based multilevel algorithm for the p-median problem. IEEE Trans. Syst. Man Cybernet. Part B (Cybernet.) **42**(4), 1187–1202 (2012)
46. T.G. Rodrigues, K. Suto, H. Nishiyama, N. Kato, Hybrid method for minimizing service delay in edge cloud computing through VM migration and transmission power control. IEEE Trans. Comput. **66**(5), 810–819 (2017)
47. L.M.S. Russo, A.P. Francisco, Quick hypervolume. IEEE Trans. Evol. Comput. **18**(4), 481–502 (2014)
48. U.K. Sikdar, A. Ekbal, S. Saha, MODE: multiobjective differential evolution for feature selection and classifier ensemble. Soft Comput. **19**, 3529–3549 (2015)
49. R. Storn, K. Price, Differential evolution – a simple and efficient heuristic for global optimization over continuous spaces. J. Glob. Optim. **11**(4), 341–359 (1997)
50. J. Sun, S. Gao, H. Dai, J. Cheng, M. Zhou, J. Wang, Bi-objective elite differential evolution algorithm for ultivalued logic networks. IEEE Trans. Cybernet. **50**(1), 233–246 (2020)
51. G. Tian, Y. Ren, M. Zhou, Dual-objective scheduling of rescue vehicles to distinguish forest fires via differential evolution and particle swarm optimization combined algorithm. IEEE Trans. Intell. Transp. Syst. **17**(11), 3009–3021 (2016)

52. H.R. Tizhoosh, Opposition-based learning: a new scheme for machine intelligence, in *International Conference on Computational Intelligence for Modelling, Control and Automation and International Conference on Intelligent Agents, Web Technologies and Internet Commerce (CIMCA-IAWTIC'06)*, vol. 1 (2005), pp. 695–701
53. T.X. Tran, D. Pompili, Joint task offloading and resource allocation for multi-server mobile-edge computing networks. IEEE Trans. Veh. Technol. **68**(1), 856–868 (2019)
54. K.H. Truong, P. Nallagownden, Z. Baharudin, D.N. Vo, A quasi-oppositional-chaotic symbiotic organisms search algorithm for global optimization problems. Appl. Soft Comput. **77**, 567–583 (2019)
55. H. Wang, Z. Wu, S. Rahnamayan, Y. Liu, M. Ventresca, Enhancing particle swarm optimization using generalized opposition-based learning. Inf. Sci. **181**(20), 4699–4714 (2011)
56. S. Wang, Y. Zhao, J. Xu, J. Yuan, C.H. Hsu, Edge server placement in mobile edge computing. J. Parallel Distrib. Comput. **127**, 160–168 (2019)
57. W.L. Wang, W.K. Li, Z. Wang, L. Li, Opposition-based multi-objective whale optimization algorithm with global grid ranking. Neurocomputing **341**, 41–59 (2019)
58. Y. Wang, X. Lin, M. Pedram, A nested two stage game-based optimization framework in mobile cloud computing system, in *2013 IEEE Seventh International Symposium on Service-Oriented System Engineering* (2013), pp. 494–502
59. P. Wu, A. Che, F. Chu, M. Zhou, An improved exact ε-constraint and cut-and-solve combined method for biobjective robust lane reservation. IEEE Trans. Intell. Transp. Syst. **16**(3), 1479–1492 (2015)
60. X. Xiao, Y. Ma, Y. Xia, M. Zhou, X. Luo, X. Wang, X. Fu, W. Wei, N. Jiang, Novel workload-aware approach to mobile user reallocation in crowded mobile edge computing environment. IEEE Trans. Intell. Transp. Syst. **23**(7), 8846–8856 (2022)
61. Z. Xu, W. Liang, W. Xu, M. Jia, S. Guo, Efficient algorithms for capacitated cloudlet placements. IEEE Trans. Parallel Distrib. Syst. **27**(10), 2866–2880 (2016)
62. S. Yang, F. Li, M. Shen, X. Chen, X. Fu, Y. Wang, Cloudlet placement and task allocation in mobile edge computing. IEEE Internet Things J. **6**(3), 5853–5863 (2019)
63. H. Yuan, M. Zhou, Profit-maximized collaborative computation offloading and resource allocation in distributed cloud and edge computing systems. IEEE Trans. Automat. Sci. Eng. **18**(3), 1277–1287 (2021)
64. P. Zhang, M. Zhou, G. Fortino, Security and trust issues in fog computing: a survey. Futur. Gener. Comput. Syst. **88**, 16–27 (2018)
65. P. Zhang, M. Zhou, X. Wang, An intelligent optimization method for optimal virtual machine allocation in cloud data centers. IEEE Trans. Automat. Sci. Eng. **17**(4), 1725–1735 (2020)
66. Q. Zhang, A. Zhou, Y. Jin, RM-MEDA: a regularity model-based multiobjective estimation of distribution algorithm. IEEE Trans. Evol. Comput. **12**(1), 41–63 (2008)
67. Z. Zhao, M. Zhou, S. Liu, Iterated greedy algorithms for flow-shop scheduling problems: a tutorial. IEEE Trans. Automat. Sci. Eng. **19**(3), 1941–1959 (2022)
68. Y. Zhou, W. Xu, Z.H. Fu, M. Zhou, Multi-neighborhood simulated annealing-based iterated local search for colored traveling salesman problems. IEEE Trans. Intell. Transp. Syst. **23**(9), 16072–16082 (2022)

Toward Secure TinyML on a Standardized AI Architecture

Muhammad Yasir Shabir, Gianluca Torta, Andrea Basso, and Ferruccio Damiani

1 Introduction

Until very recently, machine learning (ML) applications have been hosted mostly on public clouds (or possibly on high-performance on-prem systems), with organizations often relying on ready-to-go deployed models from cloud services, e.g., in many industrial applications. However, dependency on cloud-based machine learning services raises several challenges such as huge energy consumption, privacy issues, network and processing latency, and reliability issues [1, 10, 23].

TinyML is a revolutionary technology that allows ML to be run on embedded edge devices that have very limited processing power and memory. This technology has opened up a new realm of possibilities in the IoT world, enabling a variety of innovative applications such as predictive maintenance, anomaly detection, and smart monitoring. TinyML recognizes that the physical world is smarter than the existing scenario, and embedded edge devices can make decisions before seeking help from edge AI or cloud AI. This setting results in improvements such as energy efficiency, better privacy of local data processing, low processing latency, and minimal connectivity dependency.

M. Y. Shabir
University of Turin, Turin, Italy

University of Kotli AJK, Azad Jammu and Kashmir, Pakistan
e-mail: muhammadyasir.shabir@unito.it

G. Torta (✉) · F. Damiani
University of Turin, Turin, Italy
e-mail: gianluca.torta@unito.it; ferruccio.damiani@unito.it

A. Basso
Synesthesia Innovation, Turin, Italy
e-mail: andrea.basso@synesthesia.it

C. Savaglio et al. (eds.), *Device-Edge-Cloud Continuum*, Internet of Things,
https://doi.org/10.1007/978-3-031-42194-5_7

In the present work, we want to explore TinyML applications, paying particular attention to two aspects that have been given little attention in earlier studies.

First of all, we pay close attention to the TinyML security problem. We investigated several methods and approaches that are used to secure the TinyML models because they are the main challenges in the field of machine learning specially for the memory-constrained ML techniques.

Secondly, considering potential applications of TinyML (in fields such as healthcare, agriculture, education, transportation, etc.), we believe it is important to go beyond the typical model of a single ML module executed on a device. We envision TinyML applications that involve many modules, executed in a single or multiple communicating devices. A relevant initiative that aims to standardize AI-based applications is carried by the *Moving Picture, Audio, and Data Coding by Artificial Intelligence* (MPAI)[1] organization. In particular, the MPAI AIF (AI Framework) specification [19], now an IEEE standard in its version 1.1, defines a framework for designing, implementing, and deploying complex AI applications modeled as workflows of data connecting individual modules. The standard version 2.0, now under development, will cover also security functionalities of the AIF.

In the present work, starting from the state of the art, we will delineate challenges and research directions to go from simple and monolithic models to full-fledged standard-based, complex, and secure TinyML applications. The paper is structured as follows. In Sect. 2 we will describe the MPAI organization and its initiatives that are most useful for the present discussion. In Sect. 3 we will review several works from the literature, guided by our focus on TinyML applications and their security. In Sect. 4 we will outline the research directions toward our goals described above. Finally, in Sect. 5 we will give some final remarks and conclude.

2 Background on MPAI

The MPAI community is a nonprofit multinational organization with established rules, procedures, statutes, and well-organized bodies. Its main objective is to create data coding standards for AI applications.

MPAI activities include 15 standard development lines, ranging from audio to health, from autonomous cars to online gaming, and from XR to the metaverse.[2] Of particular interest in the present work is the definition of a standard *Artificial Intelligence Framework* (MPAI-AIF), which aims at facilitating the construction and automation of mixed artificial intelligence-machine learning-data processing workflows for the application areas currently under consideration by MPAI. It is worth mentioning that some authors of the present paper are members of the MPAI-AIF DC (Development Committee) and that the research proposal described

[1] https://mpai.community/.

[2] https://mpai.community/standards/.

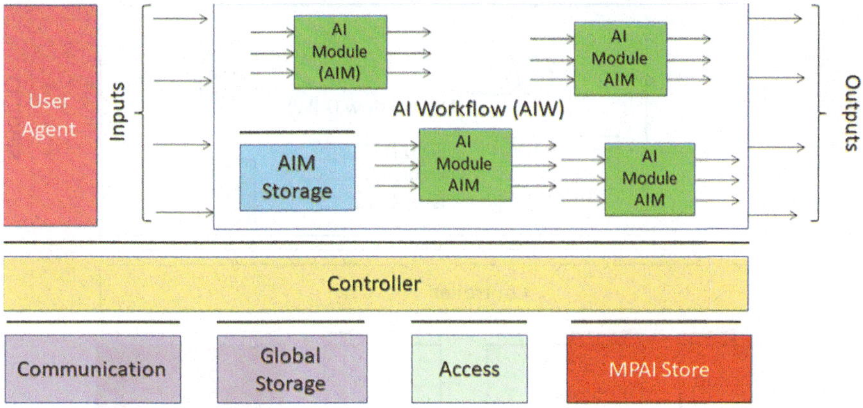

Fig. 1 Reference model of MPAI-AIF V1

in the next sections partially depends on current and future developments of the framework.

Figure 1 displays the MPAI-AIF Reference Model for V1 of the standard, in which an AIF Implementation hosts the execution of *AI Workflows* (AIW), which are made up of fundamental processing units known as *AI Modules* (AIM).

The standard architecture described by the MPAI-AIF Technical Specification [19] (adopted by IEEE as standard IEEE 3301-2022) has the following key features:

1. No dependency on specific operating system.
2. Modularity: AIWs are built on AIM components.
3. Standard interfaces encapsulate components.
4. The MPAI Store (a repository of AIWs and AIMs) gives access to verified components.
5. Components can be implemented in software, hardware, or a mix of the two (hybrid).

A prototype of the MPAI-AIF V1 has been implemented for highly constrained MCUs [2]. It is written in the C language on top of the Zephyr RTOS.

Currently, the MPAI-AIF DC is working on the definition of the second version of the standard [20], which concentrates on adding security services. Figure 2 shows the MPAI-AIF V2 Reference Model.

In particular, MPAI-AIF V2 defines a set of trusted services covering:

1. Trusted execution of AI algorithms such as NN inference (security engine)
2. Trusted handling of AI models such as (deep) neural networks (model service)
3. Attestation of the trustworthiness of HW/SW configuration (attestation service)
4. Trusted communication (communication service)
5. Trusted storage (storage service)
6. Cryptography, including encryption, signing, hashing, etc. (encryption service)

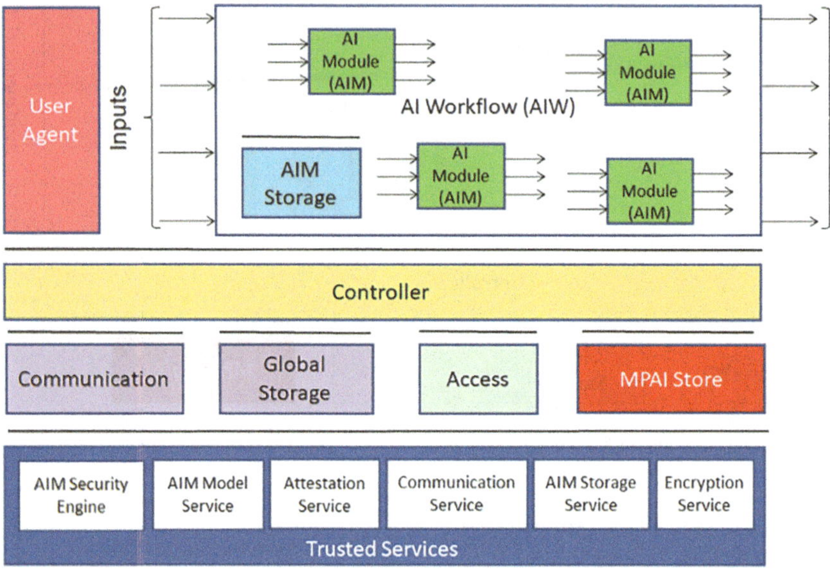

Fig. 2 Reference model of MPAI-AIF V2

3 Literature Review

In this section, we report a quite extended literature review, which is not intended as a generic survey on TinyML (such as [5]), but rather a focused selection of works in the literature that may provide the basis on which we pursue our research goals. In particular, we are interested in:

- Complex applications consisting of several cooperating modules, possibly running on several devices
- Systems that are executed exclusively (or prevalently) on constrained, far edge devices
- End-to-end security of the systems, from the single-device level to the local network-of-devices, to external communications, e.g., with the cloud

The following subsections are based on the careful review of 18 works (denoted as **P1** to **P18**) that are described and classified along dimensions that emerge from the above list of interests.

3.1 TinyML Systems

According to different studies (e.g., [11, 17, 22]), TinyML is designed to operate with a power consumption of a few milliwatts. To achieve this, the involved devices may utilize hardware acceleration and power management modules to optimize the energy efficiency of the system. The software used in TinyML must also be as compact as possible to minimize RAM requirements and power consumption. Moreover, TinyML requires optimizing machine learning models to provide better accuracy under resource constraints. When this is possible to achieve, IoT-based embedded edge devices can perform a range of tasks without relying on the cloud. In general, TinyML systems can include energy-harvesting (EH) edge devices, battery-operated embedded edge devices, sensors, and in general devices that can store just a few tens or hundreds of KB in RAM.

Our investigation of works that address constrained, intelligent devices executing ML (and AI in general) started with the following questions. Below, we'll go through each of these questions in detail. These are the following:

1. **Q1**. In the investigated system, is AI executed on end devices, edge, or both?
2. **Q2**. Is AI used for optimizing some system behavior (e.g., communication between device and edge, energy, ...)?
3. **Q3**. Is AI used for "normal" system operations (e.g., image or sound recognition)?
4. **Q4**. Is there cooperation between AI-enabled devices?

Examples of TinyML applications include smart wearables, health monitors, environmental sensors, and smart agriculture systems [24]. Smart wearables, for example, can monitor vital signs such as heart rate and detect irregularities to alert the wearer or healthcare professionals. Health monitors can be used to detect changes in patients' conditions and notify caregivers. Environmental sensors can detect changes in air quality, temperature, and humidity levels, allowing for better control of indoor environments. Smart agriculture systems can help farmers optimize crop yields by monitoring soil moisture, temperature, and other environmental factors. Table 1 reports the type of application and the problem solved by ten of the works we have revised.

Table 2 reports, for each of the papers, the answers to the four questions outlined above.

In **P1** [29], the authors identify the most challenging issue in IIoT (industrial IoT) contexts, which is ensuring a limited processing latency. The proposed framework considers heterogeneous resources and identifies the best combination of hardware to handle various AI tasks in the needed processing time. AI operations are implemented using a variety of AI libraries, such as PyTorch or TensorFlow, and the framework may link diverse AI tasks and AI hardware in a transparent way. An important issue is scheduling heterogeneous hardware for improved energy efficiency when performing various AI activities at the intelligent edge. Because unique AI hardware consumes less energy when doing specialized jobs, the proposed framework's scheduling method takes into account the task type and latency needs to minimize energy consumption while ensuring a limited processing delay.

Paper **P2** [8] presents an intelligent and secure edge-enabled computing model (ISEC) for sustainable cities using green IoT. The aim of the proposal is to improve

Table 1 TinyML applications

Paper	Application type	Solved problem
P1 [29]	Intelligent edge computing framework for AI-Drive IIoT	Scheduling problem
P2 [8]	Intelligent data transportation routing	Intelligent and secure edge-enabled computing
P3 [16]	Smart & Sustainable campus	Indoor air and temperature quality monitoring
P4 [27]	Intelligent and efficient MEC-access control	Battery prediction for mobile-edge computing
P5 [13]	Learning IoT in edge	Wireless cameras for monitoring and identifying
P6 [28]	Tasks-oriented edge computing	Reduced the bandwidth for transmission data
P7 [25]	Machine learning classifier execution	ML frameworks for classification
P8 [18]	Face mask detection technique	Quantize the CNN model
P9 [22]	Human-centric embedded machine learning	Emotion-aware facial recognition
P10 [14]	Signal recognition and channel estimation in OFDM systems	Tiny AI for channel estimation and signal detection

Table 2 TinyML systems' main characteristics

Paper	**Q1**	**Q2**	**Q3**	**Q4**
P1 [29]	Both (edge and device)	Energy optimization	Images	Yes
P2 [8]	Edge	Communication efficiency with low-energy consumption	Sensor data	No
P3 [16]	Edge	Energy optimization	Air quality and air temperature	No
P4 [27]	Server	Energy optimization	No	No
P5 [13]	Edge	No	Video sensing	Yes
P6 [28]	Both (edge and device)	Communication optimization	Images	Yes
P7 [25]	Edge and device	No	Voice, image, sound	No
P8 [18]	Device	No	Images	No
P9 [22]	Device	No	Images	No
P10 [14]	End devices	Yes	Optimize signal recognition	Yes

communication efficiency in terms of energy optimization and provide security for data transmission. The model exploits deep learning techniques for the data routing, by first predicting the best routes for the data transmission between edge servers, and then for training the sensors to make more accurate predictions for finding a transmission route. Hashing with chaining techniques is adopted to provide an efficient security solution.

In paper **P3** [16], the authors discuss the design methodology used for the architecture specification of the IPVC Smart & Sustainable Campus (IPVC-S2C), a FIWARE-based platform with edge-enabled intelligence targeting higher education institutions. The system is applied to the indoor air quality monitoring on the campus. The authors introduce the core elements and ICT infrastructure required to support the implementation of the platform while maintaining compatibility with legacy systems.

In paper **P4** [27], authors consider a dynamic MEC-access control problem in an IoT scenario that consists of k MEC (mobile-edge computing) servers and one IoT device with energy-harvesting (EH) capability. Then, they propose an intelligent long short-term memory (LSTM)-enhanced deep Q-network (DQN) access control algorithm for learning the optimal access control strategy from/to the IoT device. The learned strategy maximizes the long-term average up-link transmission rate while minimizing the transmissions' energy consumption.

In paper **P5** [13], the authors first introduce deep learning for IoT in an edge computing environment. The authors claim to have developed a novel offloading scheme to improve the performance of IoT deep learning applications with edge computing. In order to adapt deep learning models for IoT to edge computing, the authors develop an elastic model, which offloads (part of) a deep network to edge nodes. Finally, they test a deep learning model for IoT with extensive experiments in a given edge computing environment, comparing the results of their proposed edge computing method with conventional solutions.

The authors of paper **P6** [28] address the problem of effectively reducing the bandwidth needed for the transmission of data from the IoT devices to edge computing servers, such that AI model inference of IoT data can be performed. To solve the aforementioned challenge, they design TORC (tasks-oriented edge computing), a multi-task AI model inference framework for IoT edge computing with support for temporally dynamic task importance and bandwidth variation. In TORC, a neural network runs on IoT devices and encodes the data (that is intended for AI analysis tasks on the edge) into a compact representation, based on bitrate weights and importance of tasks. The output of the encoder is then quantized and compressed with an algorithm entropy coding. An edge server receives the encoded data with limited bandwidth and decodes it with the primary purpose of maintaining the accuracy of AI-based analysis tasks as high as possible.

The authors of paper **P7** [25] propose techniques that preserve the quality of the ML model regardless of the available RAM, namely, they allow fast classification using 0 bytes of SRAM. The proposed technique targets the deployment of models built with standard ML frameworks for classification, such as Python scikit-learn. To show the comparability of the results, the authors compare with known models ported with sklearn, m2cgen, and emlearn libraries and claim that the proposed technique is four times better than the competitors.

In paper **P8** [18], the authors propose a face mask detection technique applicable to smart IoT devices with very low memory. The proposed method is implemented on a tiny development board with memory constraints, such as an ARM Cortex-M7 microcontroller clocked at 480 Mhz and 496 KB frame buffer RAM. The model is quantized to further reduce its size using TensorFlow Lite framework.

In paper **P9** [22], the authors propose a TinyML technique in smart sensing devices used to detect the presence or absence of fruits in images of plants, focusing on energy consumption in various IoT scenarios. TinyML and LoRaWAN have been used in the proposed model to determine whether an energy-efficient model is capable of detecting fruit availability. The proposed model was tested for accuracy with various experiments and was shown to achieve 90% accuracy and three times energy savings on battery-powered sensors compared to similar cloud-based applications. The proposed techniques are based on the development of embedded vision AI algorithms with TinyML. This technique allows knowing when and where attention is needed by the plants, through LoRaWAN communication.

In paper **P10** [14], the authors discuss a TinyML method for hardware-efficient channel estimation and signal detection in low-cost, low-power edge devices. The method replaces large dense layers with small cascading sub-layers to reduce computation and storage requirements. The authors also introduce a rank-restricted back-propagation algorithm to facilitate fast training. The proposed techniques are validated using computer simulations of orthogonal frequency-division multiplexing (OFDM) systems using IEEE Std 802.11a-1999 and the Wireless World Initiative (WINNER II) as the channel model. The authors compare the computations and memory storage of the TinyML approach with a benchmark fully connected deep neural networks (DNN) and found that the TinyML scheme can achieve 2.5–3 times acceleration in model training and 4.5 times acceleration in model inference while reducing on-chip storage by roughly 4.5 times in comparison to the FC DNN.

3.2 Resources and Environments

One of the most significant challenges of working with TinyML is the constraints placed on resources like memory and power. In order to create applications that are both efficient and effective, researchers must develop new algorithms and methods for optimizing resource utilization. As a result, the field of TinyML research is constantly evolving, with new techniques and tools being developed all the time.

Based on our study and their analysis, we have drawn a table that shows some characteristics of the models used in the TinyML we have analyzed and of the SW/HW environments adopted. See Table 3.

3.3 Security

Secure TinyML entails the integration of a number of security measures to protect the data processed and produced by machine learning models running on

Table 3 TinyML resources and SW environments

Paper	Train	Model size	Board type or MCU	SW environment
P1 [29]	Yes	1 to 300 MB	TensorRT 7.1.3	Simulated on NS-3
P2 [8]	No	Data packet size 64 bits	ARM based 4 devices used	Tensorflow Lite
P3 [16]	No	–	ARM based 4 devices	TensorFlow Lite
P4 [27]	No	–	–	–
P5 [13]	No	Video data 95 kb/s	Simulation	Python 2.7
P6 [28]	No	Images sized 256 × 512 pixels	Simulation	PyTorch
P7 [25]	Yes	1 MB	Arduino Nano and Mega, Blue Pill, ESP 32,01s,8266	Python scikit-learn
P8 [18]	Yes	138 KB after full quantization	ARM Cortex-M7 microcontroller	TensorFlow Lite
P9 [22]	Yes	–	Smart watches and tiny personal assistants	TensorFlow Lite
P10 [14]	Yes	Proportional to the recommended compression ratios	All tiny devices	Computer simulations

microcontrollers. These techniques include secure boot techniques that ensure a secure device startup and the integrity of the software during the boot process. To secure data exchanged between the device and the cloud or other endpoints, cryptography, such as encryption and decryption algorithms, is also essential.

Access control mechanisms must be implemented by secure TinyML systems to ensure that only authorized users may access the device and data. This is enabled via user authentication techniques like passwords or biometric identification. In order to protect data transmissions and avoid data surveillance and eavesdropping, protected TinyML additionally uses secure communication protocols like Secure Socket Layer (SSL).

In a variety of fields, including healthcare, industrial IoT, and smart homes, the use of secure TinyML has become increasingly important. Examples include real-time patient monitoring and diagnosis in healthcare using TinyML models running on microcontrollers. To prevent data breaches and maintain patient privacy, it is necessary to process sensitive data, such as patient information and vital signs, with a substantial level of privacy and security.

Therefore, the implementation of secure TinyML is crucial for the adoption and deployment of TinyML solutions in various fields. The incorporation of secure boot processes, cryptography, access control, and secure communication protocols ensures the protection of sensitive data and devices from unauthorized access and cyberattacks.

The following list of security measures can be implemented to make TinyML more secure:

1. Encryption: Data, models, and firmware can be encrypted to prevent unauthorized access.
2. Authentication: Authentication protocols can be used to ensure only authorized users can access the system.
3. Secure Communication: Secure communication protocols like SSL may be utilized for secure communication between memory-constrained devices and toward external targets such as cloud servers.
4. Access Controls: Access controls can be implemented to ensure that only authorized users get access to the system.
5. Protection of User Data Privacy: Privacy approaches, including differential privacy, can be used to ensure user data privacy.

Some TinyML hardware and software security are shown in Table 4. Table 4 reports information about ten of the works we have revised that discuss security features. Two of the papers (**P2** and **P9**) have already been mentioned in the tables above, while eight papers (**P11–P18**) are specifically relevant to security.

In paper **P11** [3], the authors discuss a study that focuses on wearable devices (WD) used for healthcare monitoring. The authors categorize these technologies and identify privacy and security risks connected to their use. In addition, the authors review a number of cutting-edge technologies, such as TinyML, homomorphic encryption (HE), and federated learning (FL), that may be capable of dealing with these challenges. This study offers a taxonomy of threats and attacks that may compromise the security and privacy of WD. The study's main objectives are to emphasize the significance of protecting private health information and to propose technologies that could help reduce the associated risk. In particular, the authors

Table 4 Secure TinyML

Paper	HW security	SW security	Security techniques
P2 [8]	No	Yes	Symmetric cryptography is performed using the Diffie-Hellman algorithm
P9 [22]	No	Yes	Practical implications of adversarial threats on HC-EML applications
P11 [3]	No	Yes	Cutting-edge technologies, such as TinyML, HE, and FL
P12 [15]	No	Yes	Secure aggregation based on cryptographic schemes
P13 [12]	No	Yes	Protect the model's intellectual property using security techniques
P14 [6]	Yes	Yes	Protect the AI models developed and the training data used
P15 [26]	Yes	Yes	OTA-TinyML tested 6 ML models remotely from 4 memory units on 7 MCU boards
P16 [7]	No	Yes	EC-IG prototype: OP-TEE + QEMU on Armv8-A
P17 [9]	No	Yes	Security and privacy architecture for HioT
P18 [4]	No	Yes	Securing models: defending against three standard adversarial attacks

identify (i) the user behavior and perception, (ii) health-related data transfer, and (iii) data storage, as the three primary challenges involving security and privacy in WDs.

The authors of paper **P12** [15] discuss a research study that focuses on secure aggregation for federated learning using cryptographic schemes. The authors define the problem, categorize existing solutions, analyze challenges, and examine recent solutions of other studies. They then propose an improved definition of secure aggregation for federated learning and single out the four main privacy-enhancing technologies used to achieve secure aggregation, namely, (i) differential privacy, (ii) trusted execution environment (TEE), (iii) secure shuffling under anonymity assumptions, and (iv) cryptography.

In paper **P13** [12], authors focus on investigating other challenges that arise when edge ML is deployed and how to solve them, rather than the computational aspect. The authors explore the operational side of edge ML and discuss approaches that can be applied to various use cases, such as image recognition and natural language processing but not limited to these research areas. The paper assumes that the ML model is a deep neural network, but the discussed approaches are useful for other types of ML models as well. The authors focus on protecting the intellectual property of the ML model, particularly for TinyML and explain the challenges associated with protecting the model, providing solutions to address these challenges. The suggested techniques include model encryption, obfuscation, watermarking, and secure model training.

Paper **P14** [6] discusses the challenges faced by service providers when deploying AI models on IoT devices in a secure manner. The provider must consider the trustworthiness of the supporting platform and the range of potential threats to the device, such as direct attacks on the model IP or security of input data. To scale deployments to multiple devices and OEMs, it becomes necessary to separate untrusted code from the high-value model IP. The article emphasizes the need for secure inference outputs to prevent tampering and ensure the integrity of the AI model throughout the supply chain. The authors mention a DevSecOps workflow integrating standardized tooling for configuration, packaging, and image signing while selecting cryptographic algorithms that fit within the resources of constrained IoT devices.

In paper **P15** [26], authors propose a novel technique, called over-the-air (OTA) TinyML, which allows for the remote deployment of memory-constrained machine learning models on IoT devices. The mentioned technique enables IoT devices to perform tasks such as firmware updates, reconfiguration, and repurposing. The authors first discuss the challenges of OTA ML deployment over IoT devices from both research and engineering perspectives. Second, they claim that their OTA-TinyML framework facilitates the fetching, storage, and execution of TinyML models on resource-constrained IoT devices. It downloads the C source file of ML models from a web server to the embedded IoT devices using HTTPS. OTA-TinyML is tested by remotely fetching six types of ML models, storing them on four types of memory units, and then loading and executing them on seven popular MCU boards.

In paper **P16** [7], with the aim of enabling Industry 4.0 vertical and horizontal integration, the authors propose an edge-computing-based industrial gateway (EC-IG) for interfacing information technology and operational technology. They design and develop a working prototype to demonstrate remote production-line maintenance with a focus on security and the edge paradigm. The EC-IG brings computational resources and data storage closer to data sources, improving security aspects and enhancing the efficiency of the overall system.

The authors of paper **P17** [9] focus on the security and privacy of the Internet of Things (IoT) in healthcare applications. They highlight the challenges while implementing security frameworks and focus on the need for effective security and privacy solutions. The article aims to provide current challenges to the security and privacy of IoT in healthcare and to encourage the adoption of robust security measures to protect sensitive patient data. A security and privacy architecture for HioT is proposed.

In paper **P18** [4], the authors discuss the application of edge AI through a human-centric perspective, presenting a pipeline for developing human-centric embedded machine learning (HC-EML) applications using a generic human-centric AI (HCAI) framework. Authors also analyze the privacy, trustworthiness, robustness, and security aspects of HC-EML applications, including challenges and possible solutions. The paper includes a case study on human facial emotion recognition (FER) based on the AffectNet dataset, analyzing the effects of input quantization on the security, robustness, fairness, and trustworthiness of an EML (embedded machine learning) model. The FER model is found to be heavily influenced by certain facial features such as eyes, alar crease, lips, and jaws. Moreover, the input quantization is biased against dark skin faces, possibly due to low-contrast features.

4 Toward Complex, Secure TinyML Applications

4.1 State-of-the-Art and Research Gaps

The literature review presented in Sect. 3 has allowed us to get an understanding of the state-of-the art in technologies related to secure Tiny ML applications. From our analysis, we have drawn the following observations:

1. Many *edge-AI* systems still rely on powerful edge nodes, or even the cloud, as fundamental components to address complex AI tasks.
2. Systems that are strongly based on constrained devices (TinyML) tend to have simple architectures where:
   - end devices perform a simple ML task (e.g., image classification) individually and then
   - communicate their results to *upper-level* system components (e.g., edge servers) for storing and analyzing them.

3. Various model optimization techniques exist that make it possible to store AI models and perform inferences even in highly constrained devices.
4. Security of TinyML systems inherits methods for protecting the intellectual property (IP) of AI models from the more general ML security area (e.g., watermarking, obfuscation, cryptography).
5. Lightweight cryptographic technologies can be adopted for authentication, confidentiality, and integrity of communications, while hardware-backed trusted execution environments (TEE) can be exploited to guarantee security within single devices.

Therefore, the main gaps we have found toward our goal of enabling complex, secure TinyML applications can be summarized as follows:

1. There is currently a lack of research and implementations of complex, cooperative TinyML applications, i.e., applications composed by several modules, possibly running on separate, heterogeneous devices and exchanging information through a wireless network. Some challenges posed by such applications include:

   - *interoperability* of application modules at the data level, i.e., the data produced by a module must be consumable by other modules, possibly created by different vendors and/or reusable in different contexts;
   - seamless *connection* of modules on a single device, or multiple devices, to build an end-to-end application from sensor inputs to the final outputs; and
   - automatic *mapping* of the modules of a complex application into a network of devices, if no device is sufficiently powerful to completely host it, or partitioning is convenient for other reasons such as energy or computational efficiency.

   As we shall see, the MPAI-AIF specification [19] is a very promising starting point for addressing these challenges, although it needs to be extended in a suitable way.
2. The main gap that we could find in the state of the art of security is the lack of an application programmer-friendly framework addressing all the security aspects relevant for complex TinyML applications:

   - Portability across a variety of MCUs and RTOSes
   - Coverage of (i) single device security, (ii) communication security, (iii) intellectual property (AI models) security

   The MPAI-AIF V2 specification under development [20] can be at the same time a source and a target of the developments for addressing these open challenges (see Sect. 4.2).

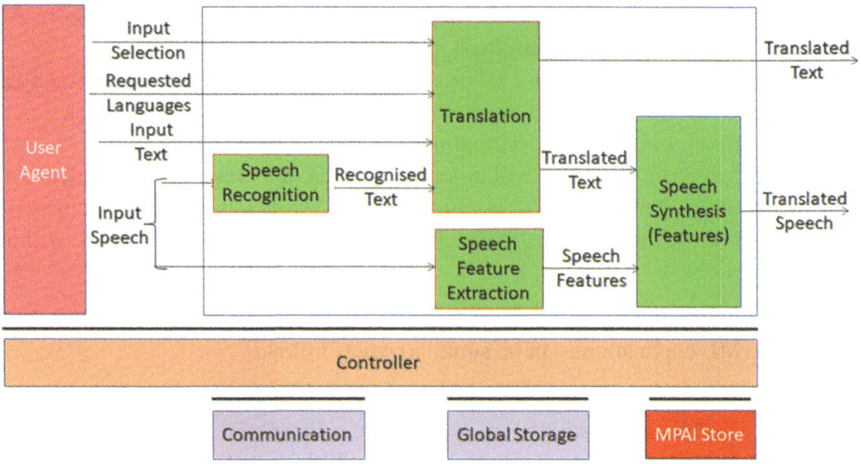

Fig. 3 MPAI MMC-UST use case

4.2 *Sketch of a Proposal*

As an example of a potential complex TinyML application, let us refer to the MMC-UST (Multi-Modal Conversation-Unidirectional Speech Translation) defined by MPAI in [21]. Figure 3 shows the workflow associated with the use case.

Without going into the details of the UC, which are beyond the scope of the present work, we note that:

- The input speech (coming from microphones) is processed by two modules, one for extracting the text (speech recognition) and one for extracting additional features, such as the intonation (speech feature extraction)
- The extracted text is processed by a translation module into a translated text
- The features and translated text are used as inputs of a speech synthesis module that produces an output audio, possibly sent to a speaker

The *data interoperability* of the modules is addressed by the MPAI MMC-UST standard, which defines, e.g., the syntax and semantics of the speech features produced by the speech feature extraction module and consumed by the speech synthesis module.

On the other hand, the MPAI specifications do not say anything about whether the modules should be hosted on a single device or *distributed* on several devices. The following is one of the many possibilities: the microphone, the speech recognition, and the speech feature extraction modules are hosted on device D_{IN}; the translation module is hosted on device D_{TR} (likely, a sufficiently powerful edge node, or even the cloud); and the speech synthesis module is hosted on device D_{OUT} which is directly connected to speakers. As already pointed out in Sect. 4.1, we need a transparent way (to the module programmer) to distribute modules across

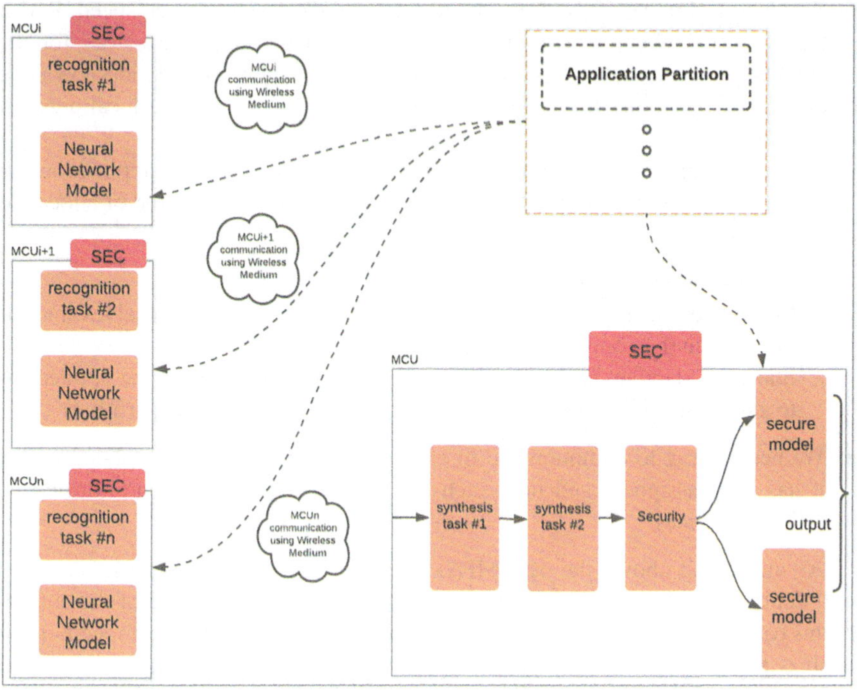

Fig. 4 Schema of a distributed TinyML application

several devices, and possibly to do this in an *automated and dynamic* way. In other words, given the *metadata* describing the modules of the use case and their data connections, we would like to automatically map the application to different devices without having to recompile its modules and, if conditions change (e.g., a communication link becomes broken or too slow), to adjust the mapping to best fit the new conditions.

In Fig. 4, we have depicted a schema of a TinyML application that is mapped to three different devices by an application partitioner.

The application partitioner could be one of the devices that will host the application, or an external node (edge, cloud). The fundamental point is that it must have access to metadata describing the application as well as to metadata describing the available devices in the local network. We can build upon and extend features offered by the MPAI-AIF to realize this architecture:

- MPAI-AIF already defines an architectural component named MPAI Store (Fig. 1), from which metadata about complete use cases (AIWs) as well as their individual components (AIMs) can be downloaded. We need to extend such metadata in order to include resource usage of modules (input and output data size and rate, RAM, flash, . . .).

- We also need to associate metadata to the network devices (clock speed, RAM, flash, sensors, wireless communications, ...); there should be a way for the application partitioner to know such metadata, either from a static knowledge base or by collecting it through some form of discovery over the local network of devices.
- We need a mapping algorithm, which decides how to partition the application and allocate its parts to different devices; many approaches for this kind of allocation are possible:
  - Optimal or approximated
  - Centralized or distributed (in the latter case, the devices themselves may be involved in the computation of the partition)
  - Static or adaptive (in the latter case, it is required to monitor the application to gather the information that may call for adjustments of the partition)
- We need a way to communicate to each device the part of the application that it has been assigned, and how such part connects with other parts hosted on different devices.

As mentioned above, the new MPAI AIF specification V2 under development will define security services offered by the AIF (Fig. 2). Since some of the authors of this paper participate to the Definition Committee (DC), the standard specification itself is going to incorporate some features required by our proposed architecture. Without committing to what will be included in the standard and what will be left (at least temporarily) outside, we can point out the following features:

- A model service will have to provide functions for the secure download and update of TinyML models (most often, but not necessarily, neural networks that have been optimized).
- A communication service will have to provide functions for the secure exchange of data between modules. The modules using this service will not know whether the module(s) they want to exchange data with are local or remote and will be determined by how the application has been partitioned.
- A firmware update service will have to provide secure update of parts of the firmware (modules of the TinyML application).
- Appropriate programmer-friendly tools will have to be available at firmware compile time to:
  - Configure the memory layout of the device distinguishing secure and nonsecure areas
  - Configure secrets (e.g., keys) to be provisioned on the device and provision them
  - Configure the firmware for including just the security algorithms actually needed by the application (and their dependencies)

5 Conclusions

The recent trend of performing machine learning tasks on memory-constrained devices at the edge, known as TinyML, offers several advantages such as lower-energy consumption and increased privacy. However, it also presents significant challenges such as performing complex ML tasks on microcontrollers with limited resources. Additionally, the security threats associated with these devices are a major concern.

In this work, we have provided a review of related literature and proposed a research direction to advance the state of the art in secure and energy-efficient ML applications, with a special focus on a standardized architecture proposed by the MPAI organization for far edge AI applications. Overall, addressing the challenges of TinyML while leveraging its benefits has the potential to transform the field of AI and bring advanced machine learning capabilities to smart devices at the edge.

Acknowledgments This publication was partially supported by the Italian PRIN project "CommonWears" (2020HCWWLP) and the EU/MUR FSE PON-R&I 2014–2020. The work is also part of the project NODES which has received funding from the MUR – M4C2 1.5 of PNRR with grant agreement no. ECS00000036. The work was also carried out within the Agritech National Research Center and received funding from the European Union Next-GenerationEU (PIANO NAZIONALE DI RIPRESA E RESILIENZA (PNRR) – MISSIONE 4 COMPONENTE 2, INVESTIMENTO 1.4 – D.D. 1032 17/06/2022, CN00000022). This manuscript reflects only the authors' views and opinions, and neither the European Union nor the European Commission can be considered responsible for them.

References

1. Y. Arfat, G. Mittone, R. Esposito, B. Cantalupo, G.M. De Ferrari, M. Aldinucci, Machine learning for cardiology. Minerva Cardiol. Angiology **70**(1), 75–91 (2022). http://doi.org/10.23736/S2724-5683.21.05709-4
2. A. Basso, D. Bortoluzzi, G. Torta, Implementation of an IoT wearable prototype on a standard ai architecture, in *2022 IEEE International Conference on Dependable, Autonomic and Secure Computing, International Conference on Pervasive Intelligence and Computing, International Conference on Cloud and Big Data Computing, International Conference on Cyber Science and Technology Congress (DASC/PiCom/CBDCom/CyberSciTech)* (IEEE, Piscataway, 2022), pp. 1–5
3. E. Boumpa, V. Tsoukas, A. Gkogkidis, G. Spathoulas, A. Kakarountas, Security and privacy concerns for healthcare wearable devices and emerging alternative approaches, in *Proceedings of the Wireless Mobile Communication and Healthcare: 10th EAI International Conference, MobiHealth 2021, Virtual Event, November 13–14, 2021* (Springer, Berlin, 2022), pp. 19–38
4. M.A. Butt, A. Qayyum, H. Ali, A. Al-Fuqaha, J. Qadir, Towards secure private and trustworthy human-centric embedded machine learning: an emotion-aware facial recognition case study. Comput. Secur. **125**, 103058 (2023)
5. D.L. Dutta, S. Bharali, Tinyml meets IoT: a comprehensive survey. Internet Things **16**, 100461 (2021). https://doi.org/10.1016/j.iot.2021.100461
6. B. Fletcher, Confidential AI for MCUs. White paper (2021)

7. S. Gupta, An edge-computing based industrial gateway for industry 4.0 using arm trustzone technology. J. Ind. Inform. Integ. **33**, 100441 (2023)
8. K. Haseeb, I.U. Din, A. Almogren, I. Ahmed, M. Guizani, Intelligent and secure edge-enabled computing model for sustainable cities using green internet of things. Sustainable Cities Soc. **68**, 102779 (2021)
9. S.M. Karunarathne, N. Saxena, M.K. Khan, Security and privacy in IoT smart healthcare. IEEE Int. Comput. **25**(4), 37–48 (2021)
10. S. Kausar, X. Huahu, W. Ahmad, M.Y. Shabir, A sentiment polarity categorization technique for online product reviews. IEEE Access **8**, 3594–3605 (2019)
11. I. Khan, A. Guerrieri, G. Spezzano, A. Vinci, Occupancy prediction in buildings: an approach leveraging LSTM and federated learning, in *2022 IEEE International Conference on Dependable, Autonomic and Secure Computing, International Conference on Pervasive Intelligence and Computing, International Conference on Cloud and Big Data Computing, International Conference on Cyber Science and Technology Congress (DASC/PiCom/CBDCom/CyberSciTech)* (IEEE, Piscataway, 2022), pp. 1–7
12. S. Leroux, P. Simoens, M. Lootus, K. Thakore, A. Sharma, Tinymlops: operational challenges for widespread edge AI adoption, in *2022 IEEE International Parallel and Distributed Processing Symposium Workshops (IPDPSW)* (IEEE, Piscataway, 2022), pp. 1003–1010
13. H. Li, K. Ota, M. Dong, Learning iot in edge: deep learning for the internet of things with edge computing. IEEE Netw. **32**(1), 96–101 (2018)
14. H. Liu, Z. Wei, H. Zhang, B. Li, C. Zhao, Tiny machine learning (tiny-ml) for efficient channel estimation and signal detection. IEEE Trans. Veh. Technol. **71**(6), 6795–6800 (2022)
15. M. Mansouri, M. Önen, W.B. Jaballah, M. Conti, Sok: secure aggregation based on cryptographic schemes for federated learning. Proc. Privacy Enhancing Technol. **1**, 140–157 (2023)
16. P. Martins, S.I. Lopes, A. Curado, Designing a fiware-based smart campus with IoT edge-enabled intelligence, in *Trends and Applications in Information Systems and Technologies*, vol. 39 (Springer, Berlin, 2021), pp. 557–569
17. M. Merenda, C. Porcaro, D. Iero, Edge machine learning for AI-enabled IoT devices: a review. Sensors **20**(9), 2533 (2020)
18. P. Mohan, A.J. Paul, A. Chirania, A tiny CNN architecture for medical face mask detection for resource-constrained endpoints, in *Innovations in Electrical and Electronic Engineering: Proceedings of ICEEE 2021* (Springer, Berlin, 2021), pp. 657–670
19. MPAI Community: Artificial intelligence framework (mpai-aif) v1.1. https://mpai.community/standards/resources/
20. MPAI Community: Artificial intelligence framework (mpai-aif) v2.0. https://mpai.community/standards/mpai-aif/about-mpai-aif/#V2, under development
21. MPAI Community: Multi-modal conversation (mpai-mmc) v1.2. https://mpai.community/standards/resources/
22. C. Nicolas, B. Naila, R.C. Amar, Tinyml smart sensor for energy saving in internet of things precision agriculture platform, in *2022 Thirteenth International Conference on Ubiquitous and Future Networks (ICUFN)* (IEEE, Piscataway, 2022), pp. 256–259
23. M. Rashid, M.A. Khan, M. Alhaisoni, S.H. Wang, S.R. Naqvi, A. Rehman, T. Saba, A sustainable deep learning framework for object recognition using multi-layers deep features fusion and selection. Sustainability **12**(12), 5037 (2020)
24. M. Shafique, T. Theocharides, V.J. Reddy, B. Murmann, TinyML: current progress, research challenges, and future roadmap, in *2021 58th ACM/IEEE Design Automation Conference (DAC)* (IEEE, Piscataway, 2021), pp. 1303–1306
25. B. Sudharsan, P. Patel, J.G. Breslin, M.I. Ali, Ultra-fast machine learning classifier execution on IoT devices without sram consumption, in *2021 IEEE International Conference on Pervasive Computing and Communications Workshops and Other Affiliated Events (PerCom Workshops)* (IEEE, Piscataway, 2021), pp. 316–319
26. B. Sudharsan, J.G. Breslin, M. Tahir, M.I. Ali, O. Rana, S. Dustdar, R. Ranjan, OTA-TinyML: over the air deployment of tinyml models and execution on IoT devices. IEEE Internet Comput. **26**(3), 69–78 (2022)

27. L. Xu, M. Qin, Q. Yang, K. Kwak, Deep reinforcement learning for dynamic access control with battery prediction for mobile-edge computing in green IoT networks, in *2019 11th International Conference on Wireless Communications and Signal Processing (WCSP)* (IEEE, Piscataway, 2019), pp. 1–6
28. J. Zhang, W. Zhang, J. Xu, Bandwidth-efficient multi-task AI inference with dynamic task importance for the internet of things in edge computing. Comput. Netw. **216**, 109262 (2022)
29. S. Zhu, K. Ota, M. Dong, Green AI for IIoT: energy efficient intelligent edge computing for industrial internet of things. IEEE Trans. Green Commun. Netw. **6**(1), 79–88 (2021)

Deep Learning Meets Smart Agriculture: Using LSTM Networks to Handle Anomalous and Missing Sensor Data in the Compute Continuum

Riccardo Cantini, Fabrizio Marozzo, and Alessio Orsino

1 Introduction

Smart agriculture, also known as precision agriculture or digital farming, is an innovative approach that utilizes advanced technologies and data-driven solutions to optimize agricultural processes, enhance crop yields, and mitigate environmental impacts [7]. With the rapid growth of the global population, the increasing challenges of climate change, resource scarcity, and changing consumer demands, there is a growing need to revolutionize traditional farming practices and adopt smarter and more sustainable methods of food production. Smart agriculture leverages cutting-edge technologies such as Internet of Things (IoT) devices, drones enhanced with swarm intelligence algorithms, artificial intelligence (AI), and big data analytics to monitor, analyze, and manage various aspects of farming operations, from soil and water management to crop monitoring and livestock health. By leveraging real-time data and advanced analytics, smart agriculture enables farmers to make informed decisions, optimize resource utilization, reduce costs, and improve overall farm productivity while also minimizing environmental impacts.

Despite the great benefits that can be obtained from these systems, their development poses a series of technological challenges, due to the collection, integration, storage, and processing of the vast amount of high-velocity heterogeneous data produced by IoT devices [2, 17]. In this context, edge computing has emerged as an effective paradigm that provides a way to overcome these challenges, by pushing part of the computation, logic, and intelligence close to the data source, i.e., the edge of the network [12]. In this way, solutions that only rely on the cloud can be improved in terms of bandwidth consumption, latency, and privacy

R. Cantini (✉) · F. Marozzo · A. Orsino
Department of Informatics, Modeling, Electronics and Systems (DIMES), University of Calabria, Arcavacata, Italy
e-mail: rcantini@dimes.unical.it; fmarozzo@dimes.unical.it; aorsino@dimes.unical.it

C. Savaglio et al. (eds.), *Device-Edge-Cloud Continuum*, Internet of Things,
https://doi.org/10.1007/978-3-031-42194-5_8

preservation, which allows for meeting the quality of service requirements even in the case of large-scale applications [3]. However, resource constraints of edge devices lead to the need for integration with cloud computing to allow heavy long-term computation and persistent storage of vast amounts of data, from which the concept of the compute continuum arises. This results in the design of complex solutions that necessitate thorough testing prior to their implementation in real-world settings, with the aim of evaluating performance, functionality, safety, security, and scalability. Simulation provides a controlled and cost-effective environment for testing and experimentation, mitigating risks associated with real-life deployment. However, challenges such as accurate modeling, availability of realistic datasets, and interoperability issues need to be addressed to ensure the reliability and validity of simulation results.

In this chapter, we demonstrate how the compute continuum can be effectively leveraged in the context of smart agriculture. The objective is to facilitate greenhouse monitoring and management while maintaining a high level of reliability and scalability. To achieve this, we incorporate a long short-term memory (LSTM)-based component into the IoT system, which effectively handles missing and anomalous sensor data. This integration enhances the productivity of the system and minimizes its environmental impact by reducing water waste and greenhouse gas emissions. A thorough experimental evaluation was performed to assess the effectiveness of the proposed solution from a twofold viewpoint. On the one hand, we evaluated the neural network accuracy in estimating the actual greenhouse state, used to replace missing values and correct anomalies. On the other hand, we leveraged the iFogSim simulation toolkit to show that deploying the proposed system at the edge-to-cloud continuum ensures higher scalability in terms of bandwidth and latency, compared to a conventional cloud-based solution. Our findings reveal the possibilities that arise from the joint utilization of the compute continuum and deep learning for sustainable and efficient agricultural practices.

The structure of this chapter is as follows. Section 2 discusses related work, highlighting the main application of deep learning to smart agriculture. Section 3 describes the proposed system, discussing how the LSTM-based component can be integrated into the greenhouse management system and how the entire system can be deployed at the edge-to-cloud continuum. Section 4 describes the experimental evaluation and simulation results. Finally, Sect. 5 concludes the chapter.

2 Related Work

This work sits at the intersection of deep learning and smart agriculture, investigating how a neural-based component, which relies on an LSTM network, can be effectively integrated into an IoT system devoted to controlling the life cycle of a smart greenhouse, supporting the decision-making process and enhancing robustness and reliability. There are many contributions in the literature that integrate deep learning architectures into IoT systems for smart agriculture, based on different neural architectures like *convolutional neural networks* (CNNs) and *recurrent neural networks* (RNNs). CNNs, inspired by the functioning of the visual cortex in the human brain, have a strong capability in image processing, which makes them widely used in agriculture research for tasks like plant or crop classification, pest detection, plant disease detection, and disaster monitoring, just to name a few [10, 16, 21]. Large pre-trained CNN architectures, such as AlexNet, are also leveraged in this context, as they can be effectively fine-tuned and applied to downstream classification tasks like disease detection and plant classification [20]. RNNs, on which this chapter is focused, are better suited to process time-series data. In the following, the main concepts behind these deep learning architectures are discussed, along with the main applications to smart agriculture.

Long Short-Term Memory Networks

Recurrent neural networks (RNNs) are a type of artificial neural network architecture that are designed to process sequential and time-series data, where the order of input elements matters. The main idea behind RNNs is to use memory to contextualize the current input based on past information. This kind of persistency, which is absent in traditional feed-forward neural networks, is achieved through recurrent connections that allow the network to store information from previous time steps. Specifically, the network maintains a *hidden state* that encodes information from what has been processed so far, thus capturing and modeling temporal dependencies in the data. *Long short-term memory* (LSTM) is a type of RNN architecture that was introduced to address the limitations of traditional RNNs, such as the *vanishing gradient* problem, which can hinder the network in learning long-term dependencies in the data. LSTMs leverage special memory cells and a gating mechanism that allow them to control the flow of information, addressing the vanishing gradient problem. This allows the network to capture and store important information over longer sequences, selectively forgetting irrelevant information. Let $W_{f,i,o}$ and $b_{f,i,o}$ be the set of weights and biases of the different gates, h_t the hidden state at time step t, x_t the input at time t, and C_t the current cell state. The LSTM includes three gates performing the following operations:

- *Forget gate* (f_t): it determines to what extent the components of the cell state must be maintained, by calculating a score using the sigmoid function:

$$f_t = \sigma(W_f \cdot [h_{t-1}, x_t] + b_f)$$

- *Input gate* (i_t): it determines what new information to store in the cell state. In particular, a sigmoid layer chooses which state values should be updated, while a tanh layer determines the new candidate values $\tilde{C}_t$:

$$i_t = \sigma(W_i \cdot [h_{t-1}, x_t] + b_i), \quad \tilde{C}_t = tanh(W_c \cdot [h_{t-1}, x_t] + b_c)$$

  At this time, the cell state can be updated as $C_t = f_t \times C_{t-1} + i_t \times \tilde{C}_t$.
- *Output gate* (o_t): it determines the final output of the module as a filtered version of the updated cell state:

$$o_t = \sigma(W_o \cdot [h_{t-1}, x_t] + b_o), \quad h_t = o_t \times tanh(C_t)$$

Main Applications to Smart Agriculture

Recurrent neural networks, along with more sophisticated variants of RNNs like long short-memory (LSTM) and gated recurrent unit (GRU) networks, have been widely used in smart agriculture. Among the main tasks, these networks were adopted for land cover classification, which involves the identification and categorization of different types of land cover classes, such as crops, vegetation, water bodies, built-up areas, and bare soil, within an agricultural field or a larger agricultural landscape using remote sensing techniques. For this kind of task, RNNs can be effectively exploited, as they can capture how land cover like the vegetation changes its spatial appearance periodically over time [6, 14]. Among the other task, RNNs were utilized for phenotype recognition, crop yield estimation, weather prediction, and soil moisture estimation [4, 11, 13, 19]. In this study, LSTM networks are leveraged for sensor data forecasting, with the aim of detecting and handling missing and anomalous sensor data. This neural component is integrated into a smart greenhouse management IoT system, allowing for the enhancement of the overall robustness and reliability, by effectively supporting decision-making and resource utilization.

3 Proposed System

This section provides an in-depth description of the proposed system, which combines an IoT system for monitoring and managing a smart greenhouse with a deep learning component based on an LSTM neural network. The role of the neural component is to support decision-making, allowing the proposed system to cope with the presence of missing and anomalous sensor data, thus improving the overall robustness and reliability of the system.

The proposed system, whose architecture is shown in Fig. 1, is composed of the following components:

- The *smart greenhouse*, which is equipped with a set of sensors and actuators. Sensors are used to measure different parameters such as temperature, humidity, light, and CO_2, which make up the current state of the greenhouse at time t, i.e., $\mathcal{S}^t$. Actuators, like fans and irrigation systems, are used to perform actions in the greenhouse in an autonomous way, based on its current state.
- The *LSTM-based component* receives data gathered by sensors, representing the current state $\mathcal{S}^t$. It is made up of two sub-components: (i) a *state regressor* which computes an estimate $\mathcal{E}^t$ of the current state, based on the previous k states, and (ii) a *state corrector* that uses the predicted state $\mathcal{E}^t$ to replace missing values of $\mathcal{S}^t$ and correct anomalies, computing the corrected state $\mathcal{C}^t$.
- The *decision-maker* receives the corrected state $\mathcal{C}$ and determines whether an actuation is needed, sending a command to the actuators installed in the

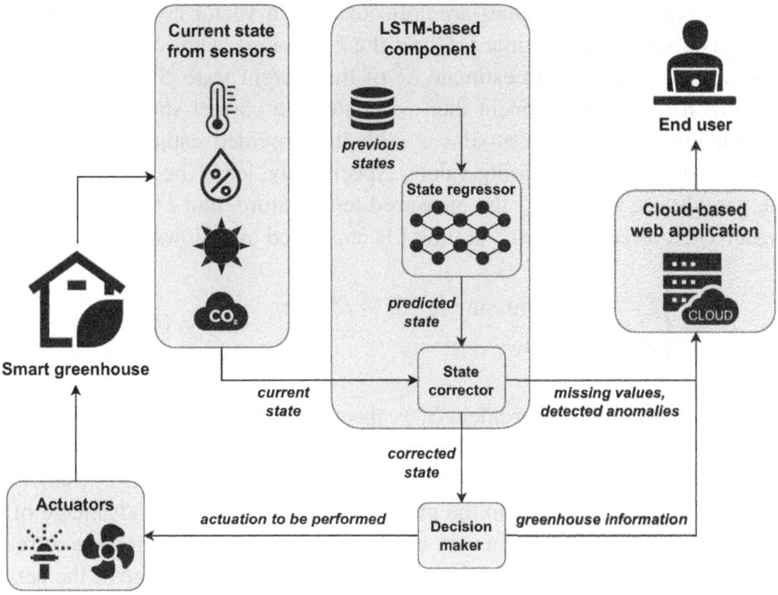

Fig. 1 Architecture of the proposed system for smart greenhouse management

greenhouse. By using the values corrected by the neural component, the system avoids being affected by an incorrect representation of the current state of the greenhouse, caused by distorted measurements related to noise and malfunctions.

- The *cloud-based web application* provides an interface to the end user, in which the current state of the greenhouse is described, allowing also to manually trigger an actuation. Moreover, all the corrections performed by the *state corrector* are sent to the web application, as they may imply the need for physical intervention. In fact, besides the presence of noise in the measurement process, an anomalous/missing value could be caused by a not working sensor –that should be physically substituted– or by an unexpected situation currently taking place.

In the following, we describe how the LSTM-based component is used for forecasting and correcting the greenhouse state measured by the IoT sensors. We also analyze how the proposed system can be deployed at the edge-to-cloud continuum to improve scalability and responsiveness, through the combined use of computing resources and capabilities at different levels of the IoT architecture.

3.1 LSTM-Based State Forecasting and Correction

IoT sensors installed in the greenhouse continuously monitor its current state, measuring different parameters including the actual temperature, humidity, light, and CO2 level. All these values are collected into a vector $\mathcal{S}^t$ that represents the current greenhouse state at time t. Given the k previous states $\mathcal{S}^{t-k}, \ldots, \mathcal{S}^{t-1}$, the *state regressor* computes an estimate $\mathcal{E}^t$ of the current state $\mathcal{S}^t$ that is sent to the *state corrector*. This component then computes the correct state $\mathcal{C}^t$ starting from the actual greenhouse state at time t with the provided estimate, by correcting anomalies and replacing missing values. Specifically, let $\mathcal{S}_i^t$ be the i-th component of the greenhouse state, e.g., the measured temperature, and $\mathcal{E}_i^t$ the corresponding estimate. The corrected state $\mathcal{C}^t$ at time t is computed as follows:

$$\mathcal{C}_i^t = \begin{cases} \mathcal{E}_i^t & \text{if } \mathcal{S}_i^t \text{ is missing or } |\mathcal{S}_i^t - \mathcal{E}_i^t| \geq \tau_i \\ \mathcal{S}_i^t & \text{otherwise} \end{cases}, \forall i \in \{1, \ldots, |\mathcal{S}^t|\}$$

In other words, if a measurement $\mathcal{S}_i^t$ is absent or differs from the prediction more than a fixed threshold τ_i, i.e., it is anomalous, it is replaced with the corresponding LSTM estimate $\mathcal{E}_i^t$. It is worth noting that a different threshold is used for the different parameters included in the greenhouse state. The main advantage of such an approach is the adoption of a very efficient and fast strategy for state correction, which is desired when running such a component close to the edge of the network, due to constrained storage and computation.

3.2 Deployment at the Edge-to-Cloud Continuum

The edge-to-cloud compute continuum provides a seamless integration of computing resources and capabilities that span from the edge of the network (i.e., the IoT devices) to the cloud, enabling data processing and analysis at different levels of the IoT architecture. This continuum encompasses three main layers.

- The *edge* layer is the closest to the source of data generation, i.e., the edge of the network. Edge devices, like IoT sensors, gateways, and edge servers, are typically smaller in scale and have limited computing capabilities. Edge computing enables to process data locally, which can result in lower latency, improved response times, and reduced bandwidth usage.
- The *fog* layer is an intermediate layer between edge and cloud, which includes resources that are placed closer to the edge of the network but are more powerful and capable than edge devices. Fog nodes can offload some of the processing tasks from edge devices, providing more advanced computing and storage capabilities compared to pure edge computing while still maintaining lower latency with respect to the cloud.
- The *cloud* layer refers to remote data centers and centralized computing infrastructures. It provides high processing power, vast storage capacity, high scalability, elasticity, and cost efficiency. This makes cloud computing ideal for handling large-scale data processing and complex analytics and for use cases that require

massive data storage, large-scale data processing, and global accessibility, such as big data analytics, artificial intelligence, and web applications.

The proposed system can benefit from a deployment on a multitier architecture like the one described above, by effectively exploiting the resources and capabilities provided by the different levels of the edge-to-cloud continuum. In such a scenario, an IoT gateway should be placed at the edge level, to collect and aggregate the raw values coming from the different sensors installed in the greenhouse and send them to the LSTM-based component. To allow low latency and a fast actuation, the LSTM-based component and the decision-maker should be placed near the edge of the network, as they are necessary to determine the needed actuation based on the current state of the greenhouse. The neural component can be deployed on an edge server or at least at the fog level, depending on the actual edge capabilities and the needs in terms of computation, to run neural network inference and state correction, and storage, to cache the previous states. It is worth noting that the inference process is far more lightweight with respect to the LSTM training, which can be periodically performed in the cloud to update and realign the model. Moreover, the correction strategy performed by the neural component is very easy and efficient, requiring just a few computing resources, which enables the possible deployment on constrained edge devices. Similarly, the decision-maker can be deployed at the edge or at least at the fog level, depending on how resource-intensive is the decision-making strategy, implemented for determining the needed actuations. Finally, the cloud-based web application should be placed in the cloud, as it requires global accessibility and data analytics capabilities, without any strict constraint in terms of latency.

4 Experimental Evaluation

In this section, we evaluate the performance of the proposed system from a twofold viewpoint. On the one hand, we assess the neural network accuracy for both regression and anomaly detection tasks, demonstrating its ability to provide accurate estimates of the actual greenhouse state. These estimates are used to provide missing values while also identifying and correcting anomalous data, whose presence can be due to sensor failures or unexpected events. On the other hand, we show how deploying the proposed system at the edge-to-cloud continuum, as described in the previous section, can lead to higher scalability in terms of bandwidth consumption and latency, compared to a conventional cloud-based solution.

4.1 LSTM Network Evaluation

To evaluate the effectiveness of the proposed system, we developed a proof of concept by training an LSTM network on a dataset containing 10 days of data

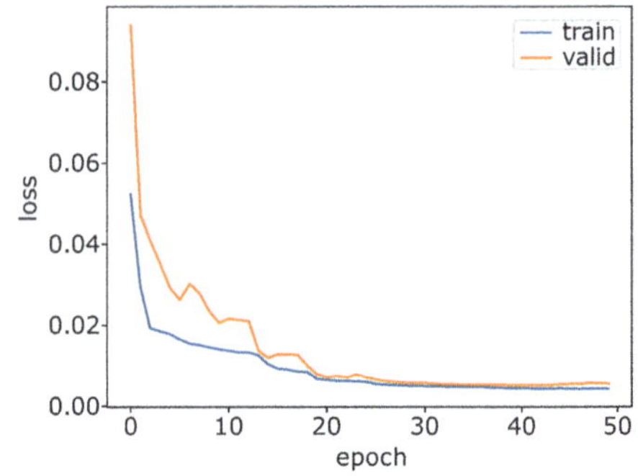

Fig. 2 Mean absolute error on train and validation set during the LSTM training

coming from four types of sensors, measuring temperature, humidity, light, and CO_2 level. The network was trained for 50 epochs with a batch size of 32, a *mean absolute error* (MAE) loss, and the *Adam* optimizer. In addition, to ensure a smoother convergence and improve performance, we scheduled the learning rate by progressively reducing its value on plateaux by a factor of 2, from an initial value of 10^{-3} to a minimum value of 10^{-5}. Learning curves are reported in Fig. 2, showing the mean absolute error on the training set and on a validation set composed of two days of measurements.

The trained regression model was also evaluated on a test set comprising four days of measurements. The results obtained show a marked ability of the model to provide very accurate estimates of the various parameters that globally constitute the state of the greenhouse, with an MAE and a coefficient of determination coefficient R^2 approaching 0 and 1, respectively. We also evaluated how precisely the estimates of the LSTM neural network approximate the individual parameters that make up the greenhouse state measured through time. The high accuracy of the model can be clearly seen in Fig. 3, which shows a comparison between the estimates of the LSTM model and the actual values included in the test set. These results are crucial to ensure the correct functioning of the entire proposed system, which relies on an accurate estimate of the current state of the greenhouse. Indeed, these estimates can be effectively exploited to replace missing values and correct any anomalies, enhancing system productivity, improving resource utilization, and reducing waste.

Once the performance of the model for the state forecasting task is assessed, we evaluated the ability of the system in identifying and correcting anomalies. For this purpose, we modified the test set by inserting random anomalies in the greenhouse state with probability p. Anomalies are added at time t to the test state $\mathcal{S}^t$ –to be predicted by the network– by multiplying it with a noise vector n sampled from a given distribution $\mathcal{D}$. Therefore, the anomalous state $\tilde{\mathcal{S}}^t$ can be represented as:

$$\tilde{\mathcal{S}}^t = \mathcal{S}_i^t \cdot n_i \ , \forall i \in \{1, \ldots, |\mathcal{S}^t|\} \ , \text{with } n \sim \mathcal{D}$$

In the performed experiments, we set $p = 0.2$, which is the probability to introduce an anomaly in the test set. Moreover, the threshold τ_i, used to detect an

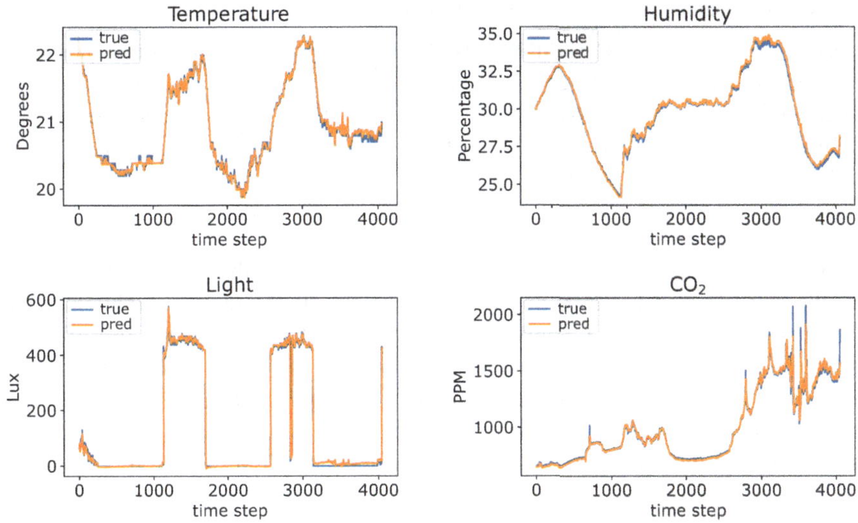

Fig. 3 Comparison between LSTM estimates and test values of each parameter included in the greenhouse state measured through time

Table 1 Anomaly detection performance for different types of anomalies, arising from uniformly and Gaussian distributed noise

	Uniform noise $n \sim U\ (a,\ b)$		Gaussian noise $n \sim \mathcal{N}\ (\mu, \sigma)$	
	$a = 1.001, b = 1.05$	$a = 1.001, b = 1.15$	$\mu = 1.002,$ $\sigma = 0.05$	$\mu = 1.002,$ $\sigma = 0.15$
Recall	0.957	0.996	0.985	0.996
Precision	0.876	0.882	0.885	0.883
f-measure	0.915	0.936	0.932	0.936

anomaly in the measurement of the i-th parameter of the greenhouse state (e.g., the temperature), was set equal to the 99-th percentile of the test MAE for that parameter. We tested the abilities of the system in the presence of different types of noise, using uniform ($n \sim U(a, b)$) and Gaussian ($n \sim \mathcal{N}(\mu, \sigma)$) distributions.

Achieved results, reported in Table 1, show the great abilities of the system in identifying true anomalies (i.e., high recall) and avoiding false positives (i.e., high precision), which results in a quite good value of f-measure, up to 0.936, in all the tested configurations.

4.2 Deployment Strategies' Comparison

In this section, we follow a simulation-based approach to investigate how the compute continuum can effectively support the proposed smart greenhouse management

References

1. A. Barbieri, F. Marozzo, C. Savaglio, IoT platforms and services configuration through parameter sweep: a simulation-based approach, in *2021 IEEE International Conference on Systems, Man, and Cybernetics (SMC)* (2021), pp. 1803–1808
2. L. Belcastro, R. Cantini, F. Marozzo, A. Orsino, D. Talia, P. Trunfio, Programming big data analysis: principles and solutions. J. Big Data **9**(4), 1–50 (2022)
3. L. Belcastro, F. Marozzo, A. Orsino, D. Talia, P. Trunfio, Edge-cloud continuum solutions for urban mobility prediction and planning. IEEE Access **11**, 38864–38874 (2023)
4. S.K. Biswas, N. Sinha, B. Purkayastha, L. Marbaniang, Weather prediction by recurrent neural network dynamics. Int. J. Intell. Eng. Inf. **2**(2–3), 166–180 (2014)
5. H. Gupta, A. Vahid Dastjerdi, S.K. Ghosh, R. Buyya, iFogSim: a toolkit for modeling and simulation of resource management techniques in the internet of things, edge and fog computing environments. Softw. Pract. Exp. **47**(9), 1275–1296 (2017)
6. D. Ienco, R. Gaetano, C. Dupaquier, P. Maurel, Land cover classification via multitemporal spatial data by deep recurrent neural networks. IEEE Geosci. Remote Sens. Lett. **14**(10), 1685–1689 (2017)
7. M.R.M. Kassim, IoT applications in smart agriculture: issues and challenges, in *2020 IEEE Conference on Open Systems (ICOS)* (IEEE, Piscataway, 2020), pp. 19–24
8. G. Kecskemeti, G. Casale, D.N. Jha, J. Lyon, R. Ranjan, Modelling and simulation challenges in internet of things. IEEE Cloud Comput. **4**(1), 62–69 (2017)
9. L.E. Lima, B.Y.L. Kimura, V. Rosset, Experimental environments for the internet of things: a review. IEEE Sens. J. **19**(9), 3203–3211 (2019)
10. H. Lu, X. Fu, C. Liu, L.g. Li, Y.x. He, N.w. Li, Cultivated land information extraction in uav imagery based on deep convolutional neural network and transfer learning. J. Mountain Sci. **14**, 731–741 (2017)
11. Z. Lu, L. Chai, S. Liu, H. Cui, Y. Zhang, L. Jiang, R. Jin, Z. Xu, Estimating time series soil moisture by applying recurrent nonlinear autoregressive neural networks to passive microwave data over the Heihe River Basin, China. Remote Sens. **9**(6), 574 (2017)
12. F. Marozzo, A. Orsino, D. Talia, P. Trunfio, Edge computing solutions for distributed machine learning, in *2022 IEEE International Conference on Dependable, Autonomic and Secure Computing, International Conference on Pervasive Intelligence and Computing, International Conference on Cloud and Big Data Computing, International Conference on Cyber Science and Technology Congress (DASC/PiCom/CBDCom/CyberSciTech)* (2022), pp. 1–8
13. D.H.T. Minh, D. Ienco, R. Gaetano, N. Lalande, E. Ndikumana, F. Osman, P. Maurel, Deep recurrent neural networks for winter vegetation quality mapping via multitemporal sar sentinel-1. IEEE Geosci. Remote Sens. Lett. **15**(3), 464–468 (2018)
14. M. Rußwurm, M. Körner, Multi-temporal land cover classification with long short-term memory neural networks. Int. Arch. Photogramm. Remote Sens. Spatial Inf. Sci. **42**, 551–558 (2017)
15. M. Sinqadu, Z.S. Shibeshi, Performance evaluation of a traffic surveillance application using iFogSim, in *International Conference on Wireless Intelligent and Distributed Environment for Communication* (Springer, Berlin, 2020), pp. 51–64
16. S. Sladojevic, M. Arsenovic, A. Anderla, D. Culibrk, D. Stefanovic, Deep neural networks based recognition of plant diseases by leaf image classification. Comput. Intell. Neurosci. **2016**, 1–11 (2016)
17. C. Sobin, A survey on architecture, protocols and challenges in IoT. Wireless Pers. Commun. **112**(3), 1383–1429 (2020)
18. C. Sonmez, A. Ozgovde, C. Ersoy, EdgeCloudSim: an environment for performance evaluation of edge computing systems. Trans. Emerg. Telecommun. Technol. **29**(11), e3493 (2018)
19. S. Taghavi Namin, M. Esmaeilzadeh, M. Najafi, T.B. Brown, J.O. Borevitz, Deep phenotyping: deep learning for temporal phenotype/genotype classification. Plant Methods **14**(1), 1–14 (2018)

20. H. Yalcin, Plant phenology recognition using deep learning: deep-pheno, in *2017 6th International Conference on Agro-Geoinformatics* (IEEE, Piscataway, 2017), pp. 1–5
21. H. Yalcin, S. Razavi, Plant classification using convolutional neural networks, in *2016 Fifth International Conference on Agro-Geoinformatics (Agro-Geoinformatics)* (IEEE, Piscataway, 2016), pp. 1–5
22. X. Zeng, S.K. Garg, P. Strazdins, P.P. Jayaraman, D. Georgakopoulos, R. Ranjan, IOTSim: a simulator for analysing IoT applications. J. Syst. Archit. **72**, 93–107 (2017)

Evaluating the Performance of a Multimodal Speaker Tracking System at the Edge-to-Cloud Continuum

Alessio Orsino, Riccardo Cantini, and Fabrizio Marozzo

1 Introduction

In recent years, the rise of the Internet of Things (IoT) has led to the generation of massive amounts of high-velocity and heterogeneous data at the network edge [2, 6]. To effectively collect, process, and analyze these data, edge-to-cloud continuum solutions have emerged, which integrate the features and services provided by both edge and cloud computing, enabling real-time and data-driven decision-making in various domains [14]. In fact, current applications for processing IoT data primarily rely on cloud computing, posing challenges related to network traffic management and response time. To address these challenges, the edge computing paradigm has been introduced, allowing for data processing closer to the data source, thus offering benefits such as low latency, privacy preservation, and scalability. Nevertheless, due to the limited resources of edge devices, there is a need to combine their capabilities with cloud computing, which allows for the persistent aggregation and resource-intensive analysis of big data.

While the edge-to-cloud compute continuum has garnered considerable attention in cutting-edge application domains such as smart cities, industrial IoT, and smart healthcare, there are still many unexplored scenarios that can benefit from it. In particular, this study focuses on speaker tracking in smart spaces, such as smart homes, offices, and public venues. The localization and tracking of speakers in these environments have become increasingly important due to the widespread use of voice assistants, security systems, and smart meeting rooms [8]. Specifically, multimodal speaker tracking, which combines audio and visual data, has been proposed as a technique to ensure accurate and robust tracking, overcoming limitations of

A. Orsino (✉) · R. Cantini · F. Marozzo
Department of Informatics, Modeling, Electronics and Systems (DIMES), University of Calabria, Arcavacata, Italy
e-mail: aorsino@dimes.unical.it; rcantini@dimes.unical.it; fmarozzo@dimes.unical.it

C. Savaglio et al. (eds.), *Device-Edge-Cloud Continuum*, Internet of Things,
https://doi.org/10.1007/978-3-031-42194-5_9

video-only or audio-only methods. In fact, video-only monitoring is limited by the camera's coverage area and faces challenges like occlusion and lighting variations, while audio-only solutions can be affected by noise and reverberations.

In this chapter, we specifically investigate the effectiveness of the edge-to-cloud continuum for multimodal speaker tracking in smart spaces, which leverages edge computing for real-time sensor data processing and cloud computing for higher-level processing and analysis. To evaluate the effectiveness of the edge-cloud integration, we followed a simulation-based approach, due to the large scale, heterogeneity, and complexity of such an IoT system, which poses significant challenges in system performance, scalability, and resource utilization. Indeed, modeling and simulation are essential to support the design and development of IoT applications, allowing a detailed evaluation before the real deployment. In particular, different design choices were investigated to understand their impact on the application performance in terms of bandwidth consumption and network latency, also comparing the performance of the edge-to-cloud continuum approach with a centralized cloud-based solution. The findings of our evaluation demonstrate the benefits of edge-cloud integration in improving application performance.

The remainder of this chapter is as follows. Section 2 provides the main concepts around multimodal speaker tracking in smart spaces and the compute continuum. Section 3 discusses related work. Section 4 describes how the multimodal speaker tracking system was modeled for simulation purposes. Section 5 presents the performance evaluation of deploying such an application at the edge-to-cloud continuum, by following a simulation-based approach. Finally, Sect. 6 concludes the chapter.

2 Background

This section provides the preliminaries to the rest of this chapter. In particular, we first discuss the task of multimodal speaker tracking in smart spaces and state-of-the-art techniques. Later, we introduce the edge-to-cloud compute continuum and its multitier structure.

Multimodal Speaker Tracking in Smart Spaces
Multimodal speaker tracking is the process of detecting and locating speakers in audiovisual contexts such as video conferences and public venues. This involves analyzing both audio and video streams to identify speakers' location and track their movements over time, which makes the task challenging due to issues like video occlusion, background noise, and changes in lighting conditions.

Generally, the audio information is used to determine the direction of arrival (DOA) of the voice, by estimating the time delay of arrival (TDOA) of the signal to different microphones. One common algorithm is the steered response power (SRP) algorithm, which analyzes the signals received by an array of microphones or sensors to determine the direction from which the sound is coming. The SRP algorithm uses a steered beamformer approach to enhance the desired signal coming from a particular direction while suppressing interference from other directions. This is done by computing the power spectral density of the received signals at different spatial locations and then steering the beamformer toward the direction with the maximum power, which corresponds to the estimated DOA of the sound source. However, a challenge with these systems is that their precision depends on the density of points to be evaluated, making them expensive for real-time applications. To solve this issue, the space to be scanned can be divided into sectors to identify possible sound sources, which can be further reduced by overlapping sectors in systems with multiple microphone arrays.

In addition to audio processing, video processing is also employed to identify potential speakers based on visual cues, such as lip movement and head orientation, and track them. These visual cues can be combined with audio-based DOA information to improve the robustness and accuracy of speaker tracking in complex environments with multiple speakers and background noise. This can be done by using the particle filter (PF) algorithm, adding the DOA information obtained from audio processing to the particle propagation stage of the PF. In particular, the PF is a type of sequential Monte Carlo (SMC) method used for estimating the state of a dynamic system based on noisy measurements. The basic idea is to represent the probability distribution of the system state using a set of discrete particles, where each particle represents a hypothesis or a guess about the true state of the system at a given time. These particles are propagated through time using a state transition model that describes how the system evolves over time. At each time step, measurements of the system are used to update the particle weights, which reflect the likelihood of each particle being the true state of the system. Therefore, particles with higher weights are considered more likely to represent the true state of the system.

Based on these concepts, different techniques have been proposed in the literature to address the problem of speaker localization and tracking in smart spaces. Qian et al. [15] proposed an algorithm that integrates audio and visual cues from a localized multimodal sensor platform, using a PF framework to dynamically combine the cues while considering audio signals measured by the maximum global coherence field (GCF). Liu et al. [10] proposed a two-layer PF algorithm for multimodal speaker tracking, generating two sets of particles from the audio and video streams independently and propagating them in separate audio and visual layers. The authors combined the audio and visual likelihoods using an adaptive sigmoid function that adjusts particle weights based on the confidence of the two modalities. In a subsequent work, the same authors [11] modified the prediction and update stage of the PF algorithm, refining the direction of the particles with multimodal information in the prediction stage and calculating the particle likelihood by combining visual distance and audiovisual direction information in the update stage. The distance likelihood was obtained using the camera projection model and the estimated size of the speaker's face, while the direction likelihood was determined by audiovisual particle fitness.

Compute Continuum

The edge-to-cloud compute continuum refers to the continuum of computing resources that span from the edge of a network to the cloud. It typically consists of the following layers, categorized based on the proximity of computing resources to the data source and the level of processing that occurs at each layer:

- The *edge layer* is the closest to the data source and generally includes edge devices, such as IoT devices, which perform local processing of the data according to the edge computing paradigm. This allows for reducing latency and bandwidth usage by processing data locally, near the data source, especially in real-time or near-real-time scenarios, such as autonomous vehicles, smart industries, smart cities, and smart spaces.
- The *fog layer* is an intermediate layer between the edge and the cloud. It includes local data centers or computing resources that are closer to the edge but are more powerful than edge devices, such as gateways. Fog computing provides edge devices with additional processing capabilities, storage, and networking resources, allowing them to offload some of the processing tasks that they are unable to complete while still maintaining lower latency than the cloud.
- The *cloud layer* refers to the centralized computing infrastructure that is typically located in remote data centers and provides on-demand computing resources, such as virtual machines, storage, and services, over the Internet. The cloud layer offers scalability, elasticity, and cost efficiency and is used for compute-intensive data processing and storage.

This distributed architecture provides several advantages and benefits that can greatly enhance the capabilities of modern computing systems. In fact, by leveraging the strengths of both edge and cloud computing, edge-cloud integration allows for real-time data processing and analysis, greater scalability, improved security and privacy, and more efficient use of resources. This can be especially useful in IoT applications such as industrial automation, autonomous vehicles, and intelligent environments like smart cities and smart spaces, where speed, reliability, and responsiveness are critical.

3 Related Work

Deploying and testing an IoT system for multimodal speaker tracking in a real-world smart space can be costly and logistically challenging. To address these issues, simulation approaches can be key to guiding design choices for IoT system modeling and validation in edge-cloud architectures. Indeed, simulation allows the exploration of different deploying strategies in the compute continuum, resource allocation policies, and system configurations in a controlled and reproducible environment. Also, it enables the evaluation of system performance under varying conditions, such as changing network conditions, workloads, and dynamic resource availability across the various layers of the edge-cloud architecture. Furthermore, it provides insights into the impact of different parameters on system behavior, helping identify performance bottlenecks and optimization opportunities. Finally, simulation is cost-effective compared to real-world implementations, as it requires no physical infrastructure and allows for the evaluation of system performance in extreme or rare scenarios that may be difficult to replicate in real-world settings [4, 7, 9].

Various open-source simulators, such as iFogSim [5], IoTSim [20], and EdgeCloudSim [19], have been proposed in the literature to simulate IoT environments, and several research works have used simulation-based approaches to test specific IoT applications on edge-cloud architectures [3, 12, 18].

4 Modeling and Simulation of a Speaker Tracking System

In this section, we describe how a speaker tracking system can be modeled and simulated using iFogSim, an open-source toolkit that provides a comprehensive framework for the modeling, simulation, and evaluation of fog computing environments. It enables the modeling of diverse elements in a fog computing environment, including fog devices, IoT devices, and cloud servers. Moreover, it facilitates the simulation of their interactions, along with the evaluation of fog computing systems' efficiency, scalability, and overall performance.

In the following of this section, we delve into the details of our study, beginning with the application modeling and simulation parameters, followed by a comprehensive overview of the edge-to-cloud architecture utilized in our simulations.

4.1 Application Modeling

The speaker tracking system comprises different modules that are connected to each other, as shown in the directed data flow (DDF) model in Fig. 1. In this model, the application is represented as a directed graph where the vertices are application modules and the directed edges convey the flow of data between modules.

In particular, the system is divided into nine modules, which are described below.

- The *sector-based detector*, which identifies when a specific sector is active, based on audio input signals. It utilizes signal processing techniques to extract relevant features and classify the sector's activity level. The information is then transmitted to the point-based locator for further processing.
- The *point-based locator*, which operates on a grid of points within the identified sector and calculates the SRP for each point. The information about the top two points with the highest SRP is then passed to the speech classifier, indicating the most probable location of the audio source.

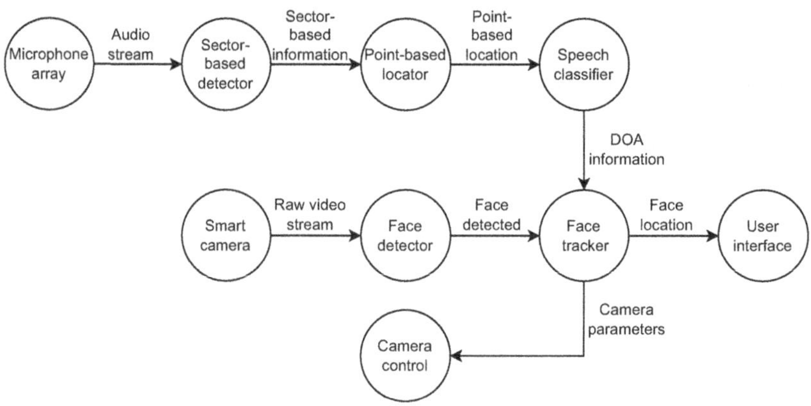

Fig. 1 DDF application model of a multimodal speaker tracking system

- The *speech classifier*, which determines if the audio response detected by the microphones originates from a human speaker. Additionally, it processes the DOA information received from the point-based locator to estimate the speaker's location. The DOA information is also forwarded to the face tracker in order to be combined with the visual information detected by the smart cameras.
- The *face detector*, which selects and transmits the frames captured by a smart camera to the face tracker.
- The *face tracker*, which employs an audiovisual particle filter (AV-PF) to accurately track the detected face's movement over time. The module continuously updates the estimated face position based on the audio and visual signals received and transmits the final estimated face position to the *user interface* for user interaction and to the *camera control* for tracking the speaker through the actuator.
- The *user interface*, which displays the frames captured by the camera and the detected face positions in real time and allows users to track the speaker's movement. It may also provide additional functionalities, such as displaying the sector information, SRP values, and DOA information for further user interaction and analysis.
- The *camera control*, which modifies the position of the smart camera to track the speaker, based on the location detected by the face tracker. This module acts as the actuator of the system.

The properties of tuples carried by edges between the modules in the application are described in Table 1, in terms of CPU length, expressed in MIPS, and network length, which refers to the communication cost between fog devices and fog nodes in the fog layer. The values were chosen according to different works of the literature [1, 5, 13, 16, 17].

Table 1 Tuple type parameters

Tuple type	CPU length (MIPS)	N/W length
Audio stream	1,000	150
Sector-based information	1,500	150
Point-based location	1,000	150
DOA estimation	2,000	100
Raw video stream	1,000	20,000
Face detected	2,000	2,000
Face location	500	2,000
Camera parameters	100	100

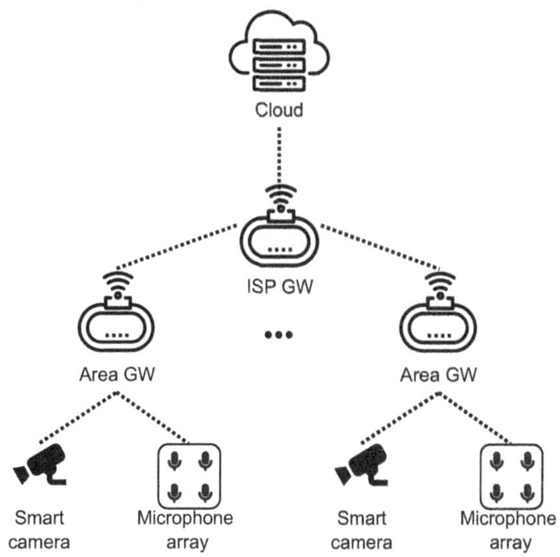

Fig. 2 Edge-to-cloud architecture to support a multimodal speaker tracking system

4.2 Edge-to-Cloud Architecture

To investigate the behavior of the system and evaluate its performance, we used an architecture composed of the following physical devices: (i) smart cameras and microphone arrays at the edge layer, which capture real-time audio and video data from the smart space; (ii) gateways at the fog layer, which support more resource-intensive tasks like speech classification or face tracking; and (iii) the cloud, which receives the processed data from the fog layer for further analysis and storage. In particular, the fog layer also handles communication with the Internet service provider (ISP) gateway, which provides access to the Internet for the cameras and microphone arrays. Moreover, the cloud layer also provides services and APIs for applications, including a user interface to interact with processed data. The system architecture is shown in Fig. 2.

Based on this architecture, we have explored two distinct strategies for the deployment and placement of application modules, namely, the *cloud-only* and the

Table 2 Configuration of the different physical devices of the architecture

Device	CPU (MIPS)	RAM (GB)	Latency (ms)	Destination
Microphone array	500	1	1	Area gateway
Smart camera	500	1	1	Area gateway
Area gateway	11,200	16	2	ISP gateway
ISP gateway	22,400	64	100	Cloud
Cloud	44,800	128	–	–

edge-ward approaches. In the former, all the modules that make up the application are deployed in a remote data center, following the traditional cloud-based deployment model. In the latter, the deployment of application modules occurs closer to the network edge. However, devices located at the network edge, such as cameras and microphone arrays, may have limited computing power, which may not be sufficient to meet the application requirements due to their resource-constrained nature. Therefore, in a such case, the fog resources are iteratively exploited up to the cloud. Table 2 describes the simulation parameters used to configure the physical topology. Specifically, the physical devices in the architecture, i.e., smart cameras, microphone arrays, area gateways, ISP gateways, and cloud –ordered by their location from edge to cloud– are described in terms of MIPS of the CPU, RAM, and the upload latency to the destination device in the architecture.

In order to investigate how the edge-to-cloud compute continuum can support the described system, we analyzed different simulation configurations with varying numbers of devices. In particular, the system was tested across various physical topology configurations with different numbers of cameras and monitored areas. We tested the use of 2 and 4 cameras and microphone arrays, combined with different numbers of covered areas, i.e., 2, 4, 8, and 16. Therefore, in each area, 2 or 4 cameras and microphone arrays can be strategically placed to monitor that area. Instead, the number of microphones was set constant to 4 throughout all simulations. In summary, a total of eight configurations, referred to as Config 1, Config 2, Config 3, Config 4, Config 5, Config 6, Config 7, and Config 8 in the experimental evaluation, have been tested as shown in Table 3. These configurations have been strategically chosen to simulate both small smart spaces that can be covered by 2 cameras and microphones, and larger spaces that need more cameras and microphones. Each simulation configuration is run according to the two deployments described above, i.e., *cloud-only* and *edge-ward*.

5 Performance Evaluation

This section presents the results of simulations carried out on an edge-to-cloud compute continuum architecture, aimed at evaluating the efficiency of two deployment strategies for the described speaker tracking system, namely, cloud-only and edge-

Table 3 Description of the different simulation configurations with a varying number of components

	Configuration							
	Config 1	Config 2	Config 3	Config 4	Config 5	Config 6	Config 7	Config 8
No. of areas	2	4	8	16	2	4	8	16
No. of cameras per area	2	2	2	2	4	4	4	4
No. of mic. arrays per area	2	2	2	2	4	4	4	4

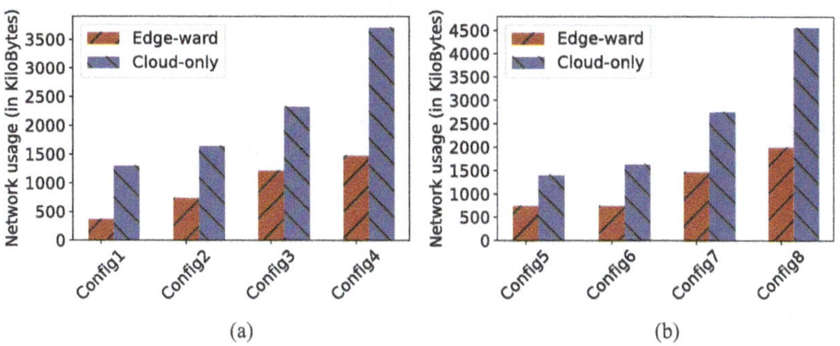

Fig. 3 Comparison of the network usage achieved by the cloud-based and edge-ward policies with eight configurations. (**a**) Small areas. (**b**) Large areas

ward. In the cloud-only strategy, all processing is done in the cloud, while in the edge-ward strategy, the processing is distributed along the edge-to-cloud continuum.

We compared the performance achieved by using the two deployment strategies in all configurations described in Table 3. The objective of these simulations was to understand to what extent an edge-cloud environment can reduce network usage and latency for a multimodal speaker tracking system.

Figure 3 shows the total network usage for each of the eight configurations. The results are similar when considering both small areas that can be monitored by 2 cameras and 2 microphones (Configs 1 to 4) and large areas that require a higher number of cameras and microphones (Configs 5 to 8). In particular, as the number of configurations becomes more complex in terms of the number of areas to be covered (from 2 to 16), the network load increases and becomes a significant challenge, especially when relying solely on cloud resources. Instead, as the results in Fig. 3 suggest, the use of fog devices can significantly reduce network usage compared to cloud-only execution. In fact, using a cloud-based approach can result in uncontrolled growth of network usage, leading to network congestion and performance degradation of the application. On the contrary, the seamless integration of computing resources along the edge-to-cloud compute continuum allows for effectively reducing the network load, mitigating the risk of network

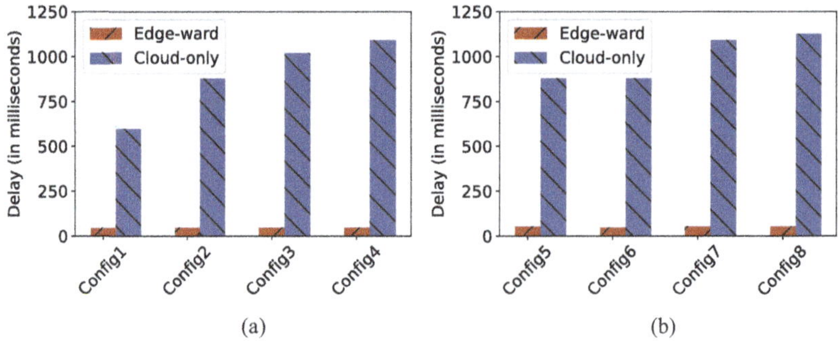

Fig. 4 Comparison of the delay achieved by the cloud-based and edge-ward policies with eight configurations. (**a**) Small areas. (**b**) Large areas

congestion, and maintaining optimal performance of the application. In fact, instead of deploying modules on the cloud, the edge-ward policy distributes modules to different locations:

- The camera control runs by default on each smart camera.
- The face detector is deployed in the cameras at the edge layer.
- The sector-based detector is deployed in the microphones at the edge layer.
- The point-based locator and the speech classifier are deployed in the area gateways.
- The face tracker is deployed in the ISP gateway.
- The user interface runs by default in the cloud.

The findings from Fig. 4 show the impact of the compute continuum on the latency of response of the system. In particular, the delays are considerably reduced when the modules are distributed over the edge-to-cloud continuum for each of the eight configurations, compared to the cloud-only deployment. This reduced delay results from the use of edge computing, where tasks are executed closer to the data source, in fog nodes or edge devices. This allows for faster processing and dramatically decreases in data transfer overheads, resulting in reduced delays.

6 Conclusions

The integration of computing resources across the edge-to-cloud compute continuum has garnered attention as a feasible solution for enabling efficient and effective collection, processing, and analysis of massive amounts of IoT data. In this chapter, we evaluated the use of an edge-to-cloud continuum architecture for a multimodal speaker tracking system, comparing two deployment strategies, namely, cloud-only and edge-ward. The simulation results reveal that leveraging computing resources along the edge-to-cloud continuum, using the edge-ward approach, consistently outperforms the cloud-only approach in all simulated configurations, especially as the number of areas to be covered by smart cameras and microphone arrays increases. Therefore, this approach enables efficient resource utilization and reduces data transfer overheads, demonstrating better overall performance in terms of latency and network usage while also ensuring robustness and scalability.

Acknowledgments This work has been supported by the "RESTART - PNRR MUR - M4C2 – I 1.3" project - CUP C37G22000480001.

References

1. M.S. Alam, S.J. Jabin, A. Alam, M.I. Hossain, Comparative analysis of cloud and Fog environment based on network usage and cost of execution using iFogSim, in *2021 International Conference on Decision Aid Sciences and Application (DASA)* (IEEE, Piscataway, 2021), pp. 132–137
2. L. Belcastro, R. Cantini, F. Marozzo, A. Orsino, D. Talia, P. Trunfio, Programming big data analysis: principles and solutions. J. Big Data **9**(4), 1–50 (2022)
3. L. Belcastro, F. Marozzo, A. Orsino, D. Talia, P. Trunfio, Edge-cloud continuum solutions for urban mobility prediction and planning. IEEE Access **11**, 38864–38874 (2023)
4. G. D'Angelo, S. Ferretti, V. Ghini, Simulation of the internet of things, in *2016 International Conference on High Performance Computing & Simulation (HPCS)* (IEEE, Piscataway, 2016), pp. 1–8
5. H. Gupta, A. Vahid Dastjerdi, S.K. Ghosh, R. Buyya, iFogSim: a toolkit for modeling and simulation of resource management techniques in the internet of things, edge and fog computing environments. Softw. Pract. Exp. **47**(9), 1275–1296 (2017)
6. N. Hassan, S. Gillani, E. Ahmed, I. Yaqoob, M. Imran, The role of edge computing in internet of things. IEEE Commun. Mag. **56**(11), 110–115 (2018)
7. G. Kecskemeti, G. Casale, D.N. Jha, J. Lyon, R. Ranjan, Modelling and simulation challenges in internet of things. IEEE Cloud Comput. **4**(1), 62–69 (2017)
8. V. Kılıç, M. Barnard, W. Wang, A. Hilton, J. Kittler, Mean-shift and sparse sampling-based SMC-PHD filtering for audio informed visual speaker tracking. IEEE Trans. Multimedia **18**(12), 2417–2431 (2016)
9. L.E. Lima, B.Y.L. Kimura, V. Rosset, Experimental environments for the internet of things: a review. IEEE Sens. J. **19**(9), 3203–3211 (2019)
10. H. Liu, Y. Li, B. Yang, 3d audio-visual speaker tracking with a two-layer particle filter, in *2019 IEEE International Conference on Image Processing (ICIP)* (IEEE, Piscataway, 2019), pp. 1955–1959
11. H. Liu, Y. Sun, Y. Li, B. Yang, 3d audio-visual speaker tracking with a novel particle filter, in *2020 25th International Conference on Pattern Recognition* (IEEE, Piscataway, 2021), pp. 7343–7348
12. S. Maheshwari, D. Raychaudhuri, I. Seskar, F. Bronzino, Scalability and performance evaluation of edge cloud systems for latency constrained applications, in *2018 IEEE/ACM Symposium on Edge Computing (SEC)* (IEEE, Piscataway, 2018), pp. 286–299
13. R. Mahmud, S. Pallewatta, M. Goudarzi, R. Buyya, iFogSim2: an extended iFogSim simulator for mobility, clustering, and microservice management in edge and fog computing environments. J. Syst. Softw. **190**, 111351 (2022)
14. F. Marozzo, A. Orsino, D. Talia, P. Trunfio, Edge computing solutions for distributed machine learning, in *2022 IEEE International Conference on Dependable, Autonomic and Secure Computing, International Conference on Pervasive Intelligence and Computing, International*

Conference on Cloud and Big Data Computing, International Conference on Cyber Science and Technology Congress (2022), pp. 1–8
15. X. Qian, A. Brutti, M. Omologo, A. Cavallaro, 3d audio-visual speaker tracking with an adaptive particle filter, in *2017 IEEE International Conference on Acoustics, Speech and Signal Processing (ICASSP)* (IEEE, Piscataway, 2017), pp. 2896–2900
16. F.H. Rahman, T.W. Au, S. Shah Newaz, W.S. Haji Suhaili, A performance study of high-end fog and fog cluster in iFogSim, in *Computational Intelligence in Information Systems: Proceedings of the Computational Intelligence in Information Systems Conference (CIIS 2018)*, vol. 3 (Springer, Berlin, 2019), pp. 87–96
17. D.M.A.d. Silva, G. Asaamoning, H. Orrillo, R.C. Sofia, P.M. Mendes, An analysis of fog computing data placement algorithms, in *Proceedings of the 16th EAI International Conference on Mobile and Ubiquitous Systems: Computing, Networking and Services* (2019), pp. 527–534
18. M. Sinqadu, Z.S. Shibeshi, Performance evaluation of a traffic surveillance application using iFogSim, in *International Conference on Wireless Intelligent and Distributed Environment for Communication* (Springer, Berlin, 2020), pp. 51–64
19. C. Sonmez, A. Ozgovde, C. Ersoy, EdgeCloudSim: an environment for performance evaluation of edge computing systems. Trans. Emerg. Telecommun. Technol. **29**(11), e3493 (2018)
20. X. Zeng, S.K. Garg, P. Strazdins, P.P. Jayaraman, D. Georgakopoulos, R. Ranjan, IOTSim: a simulator for analysing IoT applications. J. Syst. Archit. **72**, 93–107 (2017)

A Deep Reinforcement Learning Strategy for Intelligent Transportation Systems

Francesco Giannini, Giuseppe Franzè, Giancarlo Fortino, and Francesco Pupo

1 Introduction

In the last few decades, advances in vehicular networking, communication, and computing technologies have facilitated the practical deployment of autonomous vehicles (AVs). On the other hand, traffic congestion has become a serious problem and has led to an urgent need for innovative solutions to deal with it. More specifically, Internet of Vehicles (IoV) [1–3] is spreading across all the high-tech roads and will soon become the technological paradigm of assisted and autonomous driving.

In this context, Vehicle Routing Problem (VRP) [4] is a well-studied topic and several solutions have been proposed: starting from classical techniques based on optimization theory [5, 6] to the latest solutions based on Game Theory [7] and Machine Learning [8–10]. The main aim of our work is to propose new insights in the field of path planning for constrained autonomous vehicles moving in the cluttered environments of urban road networks. The application paradigm is sub-microscopic traffic simulation [11]: each vehicle and its dynamics are modeled individually. Therefore we deal with the problem of routing and controlling the vehicles while maintaining a platoon formation among them. Studies along these lines have been presented in [12, 13], and [14]. The context is an Urban Road Network (URN), with its real traffic demand [15].

F. Giannini · G. Fortino · F. Pupo (✉)
DIMES, Università della Calabria, Rende (CS), Italy
e-mail: francesco.giannini@dimes.unical.it; giancarlo.fortino@unical.it; francesco.pupo@unical.it

G. Franzè
DIMEG, Università della Calabria, Rende (CS), Italy
e-mail: giuseppe.franze@unical.it

C. Savaglio et al. (eds.), *Device-Edge-Cloud Continuum*, Internet of Things,
https://doi.org/10.1007/978-3-031-42194-5_10

The coordination and control of multi-vehicle systems have attracted the attention of several researchers in the last years, as testified by relevant review contributions, see, e.g., [16]. Model Predictive Control (MPC) seems to be the most adequate approach for dealing with this class of problems in virtue of its intrinsic capability to explicitly comply with hard constraints. In particular, in [17] a Distributed MPC strategy for solving the formation problem of leader–follower (LF) systems is derived by exploiting explicit robust model-based control techniques with set-theoretic ideas.

Additionally, the overall control architecture was designed to be flexible, scalable, and computationally efficient to be able to address operational scenarios that vary over time along the way.

Increasing the level of autonomy for vehicles in Intelligent Transportation Systems gives rise to several advantages: time savings for drivers, energy savings for the environment, and road safety improvement. In particular, travel time savings can be achieved by coordinated and connected vehicle configurations that can be more efficiently managed by means of ad hoc control strategies for autonomous multi-vehicles systems.

We use Reinforcement Learning (RL) [18] to deal with traffic congestion and to lead the autonomous vehicles to their own destinations. RL learns from experience and can reconsider its own beliefs about how to drive over the URN, so it always tries to improve what it has learned. More specifically, our solution is based on Deep Q Learning [19], and therefore a traffic evolution model is not required and the Q-function is a Neural Network.

In order to simulate the proposed DQL Algorithm in complex environments, we have developed a co-design between **S**imulation of **U**rban **MO**bility ®(SUMO) software [20] and *MATLAB 2022a*®.

Although some studies have been published on the integration between machine learning and distributed model predictive control approaches, see, e.g., [21–23] and the references therein, these topics are worth looking into the capability of the distributed MPC to take fully advantage of the potentialities of the class of RL algorithms in terms of achievable control performance.

2 Notation and Definitions

The vehicles drive within an URN, consisting of M links and F junctions:

Definition 1 A link is defined as a portion of a road between two junctions. □

Definition 2 A junction is a point over the URN where a routing action is applied. □

Definition 3 Given a road link j, at each time instant t the vehicle density $\rho^j(t)$ is defined as follows:

$$\rho^j(t) := \frac{\mu^j(t)}{\bar{d}^j l^j} \tag{1}$$

where $\mu^j(t) \in \mathbb{Z}^+$ is the number of vehicles on j at t, $\bar{d}^j \in \mathbb{R}^+$ the average road width, and l^j the road length. □

3 Problem Formulation

Let us consider an URN. The vehicle has to choose an action when it enters in the "Decision Zone" (DZ, see Fig. 1), i.e., when it is near to the next junction. "Near" can be imposed, specifically, by a constant distance or through a function of the actual vehicle longitudinal velocity. This decision zone has to be formally defined for each link, and this translates into the following half-space description:

$$DZ^j : H^j p > g^j, p \in \mathbb{R}^2, \ j = 1, \ldots, M \tag{2}$$

where H^j and g^j identify a plane in $\mathbb{R}^2$ corresponding to the decision zone on the j-link. See the explanatory example of Fig. 1.

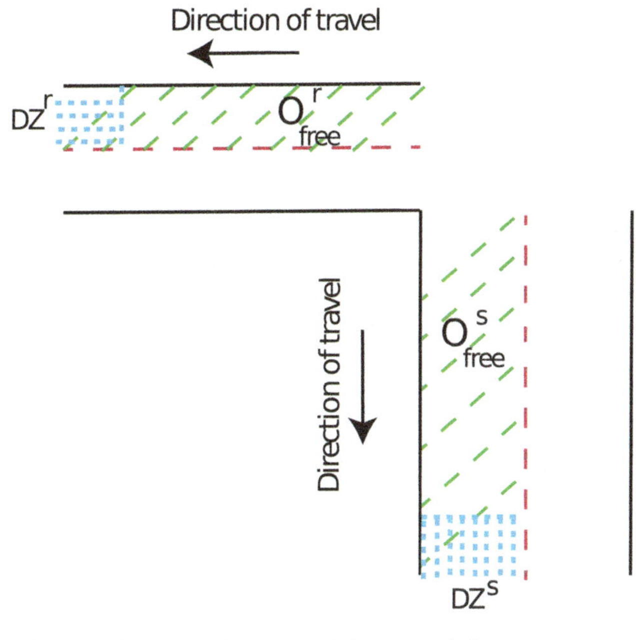

Fig. 1 Geometric constraints (dashed green zone) due to travel direction and center line (dashed red line). O^j_{free} is the obstacle-free region on the j-th road

Fig. 2 Selected views on the A. Costa scenario from iTETRIS [15]

By exploiting the decision zone as a threshold, the VRP has been recast to a path search problem on a fully connected graph. For the sake of computational complexity savings, the resulting graph can be appropriately pruned to reduce as much as possible the research space. For example, the district illustrated in Fig. 2 has been translated into a graph and pruned, as illustrated in Fig. 4.

Operating Scenario Our autonomous vehicles operate within an URN, and the decision-making process focuses on the actual traffic conditions, so the purpose of alleviating traffic congestion is achieved by adjusting the vehicles' route selection during the travel. □

Therefore, given an URN, the problem we want to solve can be stated as follows: **Autonomous Vehicles in Urban Environments (AV-UE)**: *Given a team of autonomous vehicles, find the best route in order to drive them to their destination and to avoid as much as possible traffic congestion.* □

In order to solve the **AV-UE** problem, we made the following assumptions:

- There exists an admissible path between any starting and destination pair along the URN.
- Each vehicle on the URN is connected and can communicate with another vehicle (V2V communication), with the intelligent Road Side Unit (V2R), with the cloud (V2C), or with the other connected things, e.g., traffic lights, bikes, or pedestrians (V2X).
- The environment informs the vehicle about the vehicle densities over the URN edges (by means of a server) and gives them the reward (as a broadcast message).
- The vehicles are equipped with a perception module capable of detecting obstacles within a pre-specified radius $R > 0$. Let R^c_{min} be the minimum curvature radius of the given vehicle, and then the perception module is such that the field of view is 360° and $R > R^c_{min}$. In this respect, the ball centered at the current agent planar position $\mathcal{B}(p^i(t), R)$, $\forall i = 1, \ldots, L$, represents the detected region.

3.1 Proposed Solution: An Overview

The problem has been addressed by means of a distributed control scheme properly customized to exploit the capabilities of the presented Deep Reinforcement Learning algorithm. In particular, the idea we have developed consists in exploiting routing decisions, provided by the DRL unit at each junction occurrence, as set-point references for the underlying controllers, see Fig. 3. Essentially, it consists of three units: a **Path Planner** based on the **DRL** algorithm, a **Reference Generator**, and a **Controller**, each one tuned w.r.t. the involved vehicles AV^i, $i = 1, \ldots, L$.

At each time instant t, the **Path Planner** receives the measurements of the current position in terms of coordinates and corresponding edge and the vehicles' density on that edge, from the generic vehicle AV^i at t, and generates the action

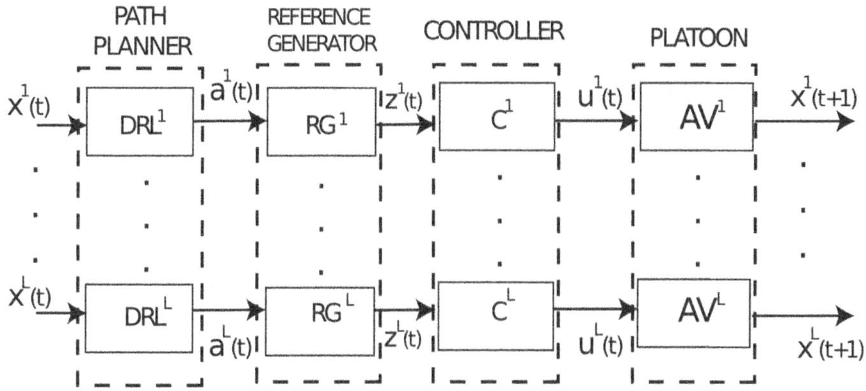

Fig. 3 Distributed control architecture exploiting the DRL algorithm

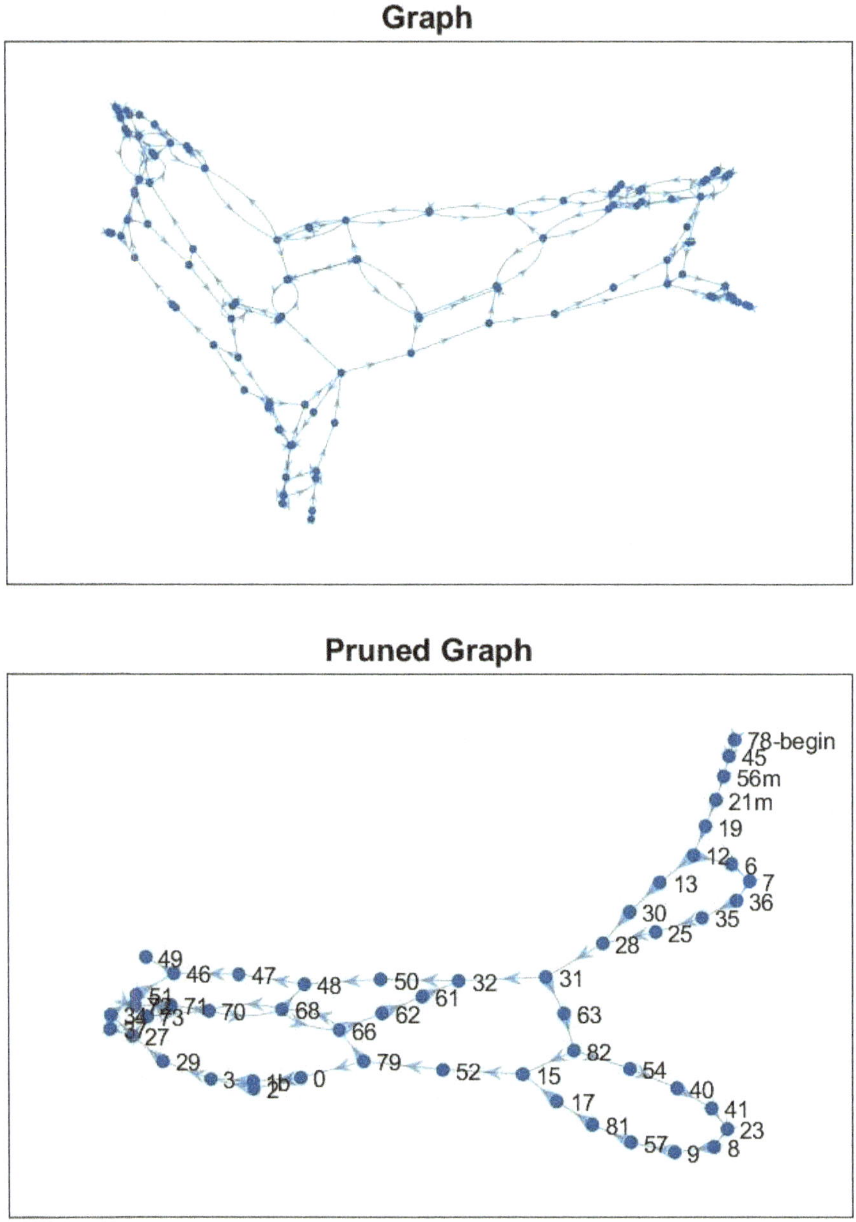

Fig. 4 The graph corresponding to the considered map (illustrated in Fig. 2) and its pruned version

$a^i(t)$ to be executed by the vehicle AV^i at the next junction. The latter translates into the set-point $z^i(t)$ for the underlying control logic thanks to the **Reference Generator** unit. Hence, the **Controller** computes an admissible command $u^i(t)$ in a distributed receding horizon fashion.

4 Reinforcement Learning Strategy

This section is devoted to implement a DRL strategy for routing decision purposes. By referring to a distributed framework, a Q learning approach is adopted and the multi-agent RL model is the following:

$$< V, \Sigma, \Lambda, \Phi > \tag{3}$$

where

- $V = \{AV^1, \ldots, AV^h\}$ accounts for the set of vehicles (agents) contributing to the RL task.
- $\Sigma := \{\sigma^i\}_{i=1}^h$ is the set of usable edges and related distribution densities $\rho^j(t),\ j = 1, \ldots, M$.
- Λ collects the admissible vehicle actions, e.g., "turn right," "turn left," and "straight" (each possible action is labeled with an integer number)
- $\Phi : \Sigma \times \Lambda \times \Sigma \to \mathbb{R}$ is the global reward function.

Notice that the RL is activated whenever the vehicle AV^i enters the decision zone. From now on, let t_{dec} denote this trigger time. The Q function $Q^i : \Sigma \times \Lambda \to \mathbb{R}$ is expressed as a Neural Network (NN), and it is usually indicated as $Q^i(\sigma^i, a^i; \theta^i)$, where θ^i indicates the NN weights. The designed NN is illustrated in Fig. 5. The blue blocks are the inputs (the upper takes σ^i and the other a^i). The two branches contain a Fully Connected Layer, an ELU activation function, and another Fully Connected Layer. The two branches are added and the output entries in an ELU and in the last Fully Connected Layer. In the blocks in Fig. 5, for each block, the size of its output is indicated. Notice that the last block is to calculate the Q value, and therefore it is reduced to just one neuron.

In general, the Q-function is defined as follows:

$$Q^i(\sigma^i(t_{dec}), a(t_{dec}); \theta^i) \triangleq \\ E\left[\sum_{\nu=1}^{\infty} \gamma^{(\nu-1)} \phi^i(t_{dec} + \nu \Delta t_{dec}(t))\right] \tag{4}$$

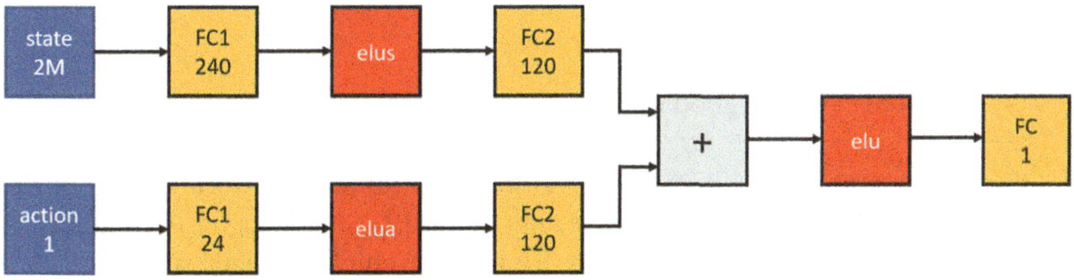

Fig. 5 Neural Network architecture

where the $E[\cdot]$ is the expected value operator, $a(t_{dec}) \in \Lambda$, $\theta^i \in \mathbb{R}^W$ collects all the NN weights, $\gamma \in (0, 1)$ is the discount factor, and ϕ^i is the local reward. It is worth to underline that the Q-function is initialized with random weights under the operating assumption that no information is available on the traffic behavior before the NN training. Moreover, the greedy action $a^i(t_{dec})$ should be chosen such that

$$a^i(t_{dec}) := \arg\max_{a\in\Lambda} Q^i(\sigma^i(t_{dec}), a; \theta^i) \tag{5}$$

During the training procedure, a trade-off between exploration and exploitation comes out. Hence, such a hitch can be overcome by using an epsilon-greedy policy, i.e., the probability $\epsilon(\cdot)$ of choosing a random action is an inversely proportional function of the time.

Here, the following local reward function has been selected:

$$\begin{aligned}\phi^i(t_{dec} + \Delta t_{dec}(t)) = \Phi\left(\bigcup_k \sigma^k(t_{dec} + \Delta t_{dec}(t))\right)\\ +\omega^i(t_{dec} + \Delta t_{dec}(t)),\, i \in V\end{aligned} \tag{6}$$

where $\Delta t_{dec}(t) := t - t_{dec}$ with $x^i(t_{dec}) \in (\mathcal{X}_c)^i_r$ and $x^i(t) \in (\mathcal{X}_c)^i_s, r \neq s$. Roughly speaking, if the action was "turn right," the value of (6) can be correctly exploited once the vehicle AV^i has effectively made such an action. Notice that the two components of (6) have the following meaning:

- The first term $\Phi\left(\bigcup_k \sigma^k(t_{dec} + \Delta t_{dec}(t))\right)$ weights the traffic congestion over the whole URN and it is computed by the following formula:

$$\begin{aligned}&\Phi\left(\bigcup_k \sigma^k(t_{dec} + \Delta t_{dec}(t))\right)\\ &= \sum_{j=1}^{M} \rho^j(t_{dec}) - \rho^j(t_{dec} + \Delta t_{dec}(t))\\ &= \sum_{j=1}^{M} \frac{\mu^j(t_{dec}) - \mu^j(t_{dec} + \Delta t_{dec}(t))}{\overline{d}^j l^j}\end{aligned} \tag{7}$$

  Notice that one obtains good rewards (performance) as long as the vehicle density decreases.
- The component $\omega^i(t_{dec} + \Delta t_{dec}(t))$ is the total cumulative waiting time of the vehicle AV^i from the initial time instant t_0 until $t_{dec} + \Delta t_{dec}(t)$. Essentially, it penalizes those sequences of actions driving the vehicle to queue scenarios.

In order to adequately train the considered NN, the following squared loss function is used:

$$\begin{aligned}&\mathcal{L}(t_{dec} + \Delta t_{dec}(t); \theta^i) := \\&(\phi^i(t_{dec} + \Delta t_{dec}(t) - \hat{\phi}^i(t_{dec} + \Delta t_{dec}(t)))^2\end{aligned} \tag{8}$$

where the expected reward is

$$\begin{aligned}\hat{\phi}^i(t_{dec} + \Delta t_{dec}(t)) = Q^i(\sigma^i(t_{dec}, a_k^i; \theta^i)-\\-\gamma \max_{a\in\Lambda} Q^i(t_{dec} + \Delta t_{dec}(t), a; \theta^i)\end{aligned} \tag{9}$$

Proof

$$\begin{aligned}&Q^i(\sigma^i(t_{dec}), a(t_{dec}); \theta^i) \triangleq \\&E\left[\sum_{\nu=1}^{\infty} \gamma^{(\nu-1)} \phi^i(t_{dec} + \nu\Delta t_{dec}(t))\right]\\&= \phi^i(t_{dec} + \Delta t_{dec}(t)) + E\left[\sum_{\nu=2}^{\infty} \gamma^{(\nu-1)} \phi^i(t_{dec} + \nu\Delta t_{dec}(t))\right]\\&= \phi^i(t_{dec} + \Delta t_{dec}(t))\\&+\gamma Q^i(\sigma^i(t_{dec} + \Delta t_{dec}(t)), a(t_{dec} + \Delta t_{dec}(t)); \theta^i) =\\&= \phi^i(t_{dec} + \Delta t_{dec}(t)) + \gamma \max_{a\in\Lambda} Q^i(t_{dec} + \Delta t_{dec}(t), a; \theta^i)\end{aligned}$$

Finally, whenever a new reward measurement is useful, a gradient descent update is performed:

$$\theta^i = \theta^i - \alpha \nabla_\theta \mathcal{L}(\theta^i) \tag{10}$$

with $\alpha \in (0, 1)$ the learning rate.
All the above developments allow to write down the following computable algorithm.

DQL Algorithm

Initialization

Extrapolate $\sigma^i(0), i = 1, \ldots, h$, from the URN;
Initialize $Q(\sigma^i, a; \theta), i = 1, \ldots, h$;

At each t_{dec} *do*

Select an admissible action:

$$a^i(t_{dec}) = \begin{cases} rand(\Lambda), \text{ with probability } \epsilon(t_{dec}) \\ \arg\max_{a\in\Lambda} Q^i(\sigma^i(t_{dec}), a; \theta^i), \text{ otherwise}\end{cases}$$

Observe the global reward $\Phi\left(\bigcup_k \sigma^k(t_{dec} + \Delta t_{dec}(t))\right)$ and $\sigma^i(t_{dec} + \Delta T_{dec}(t)), \forall i = 1, \ldots, h$;
Compute the local reward $\phi^i(t_{dec} + \Delta T_{dec}(t)), , \forall i = 1, \ldots, h$;
Update $\theta^i, \forall i = 1, \ldots, h$.

---

5 SUMO *versus* MATLAB Co-design

A co-design procedure is outlined in order to allow information sharing between SUMO and MATLAB. Essentially this prescribes that starting from the set of functions provided by TraCI4Matlab® [24] that are customized to comply with the **DQL Algorithm**, the co-design in terms of software communication and data exchange is performed. More in detail, the TraCI4Matlab toolbox allows the communication between any MATLAB application and the SUMO traffic simulator. TraCI application level protocol is based on the client–server paradigm, and it is built on top of the TCP/IP stack: the application developed in MATLAB plays the role of a client, and it can access and modify the simulation environment provided by SUMO that acts as a server. Hence, the proposed co-design is in charge of automatically mapping all the URN information into properly encoded MATLAB objects. Then, a MATLAB graph is obtained by exploiting the SUMO xml files, where the URN junctions (i.e., crossroads, traffic circles, road entrances, and exits) are the graph nodes, while the roads are the graph edges, see the explicative Fig. 4. A set of Application Programmable Interfaces (APIs) have been then developed to get, modify, or add information about vehicles within the simulation environment. For instance, these routines can be used to query SUMO where a vehicle is currently moving or to obtain data regarding road vehicles density, spatial constraints (i.e., road width, length, and topology), obstacles, and so on.

As the **DQL Algorithm** is concerned, the proposed co-design adapts to new operating scenarios by simply selecting a new SUMO xml describing file. In particular, the designed APIs allow to compute the local reward (6), while the optimization of NN weights is achieved by means of the MATLAB Reinforcement Learning toolbox® [25] and the MATLAB Deep Learning toolbox® [26].

6 Simulations

This section aims at highlighting the performance of the proposed strategy and the obtained results in a numerical scenario built on real traffic data [15].

Some simulations have been conducted by means of a platoon of ten autonomous vehicles, inserted in the proposed framework in the southwest road in the map and imposing them to reach a target location in the southeast part, see Fig. 6. More precisely, choosing a reference initial position at latitude $44°29'23.6''N$ and longitude $11°18'30.2''E$, the AV initial positions in a local tangent plane ENU reference frame [27] have been set as described in Table 1, with zero initial velocities.

The Neural Network architecture described in Fig. 5 has been developed in the Matlab Deep Network Designer, see Fig. 7.

More precisely, each agent is modeled as an ideal point that moves on a planar environment according to

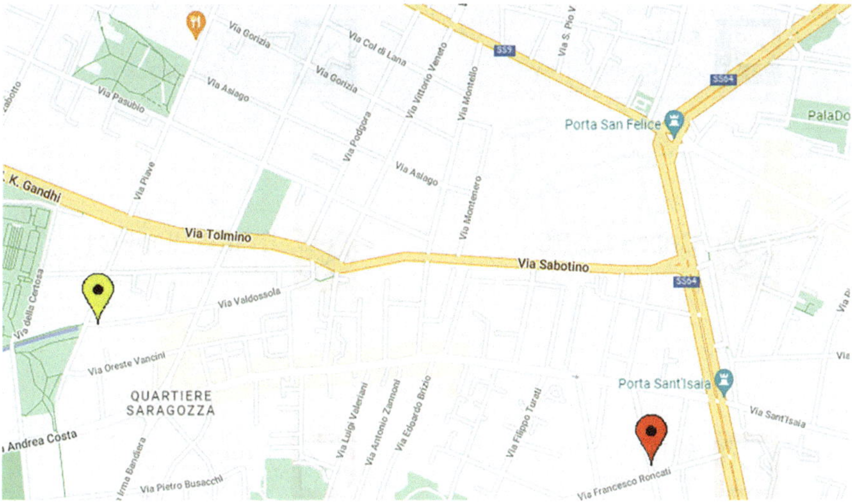

Fig. 6 SUMO URN representation with starting and target locations

Table 1 Platoon initial positions

Agent	AV^1	AV^2	AV^3	AV^4	AV^5	AV^6	AV^7	AV^8	AV^9	AV^{10}
$x_1^i(0)$	70.02	64.00	57.90	51.83	45.76	39.70	33.63	27.57	21.50	15.44
$x_2^i(0)$	386.42	384.08	381.74	379.40	377.06	374.72	372.39	370.05	367.71	365.37

$$x^i(t+1) = A^i x^i(t) + B^i u^i(t) \tag{11}$$

with the state vector $x^i(t) = [p^i(t)^T,\ \dot{p}^i(t)^T]^T \in \mathbb{R}^4$, accounting for agent planar position and velocity components, respectively, $u_i(t) = [u_1^i(t),\ u_2^i(t)]^T \in \mathbb{R}^2$ the acceleration vector and

$$A^i = \begin{bmatrix} I_2 & T_s I_2 \\ 0_2 & I_2 \end{bmatrix},\ B^i = \begin{bmatrix} \frac{T_s^2}{2} I_2 \\ T s I_2 \end{bmatrix},\ \forall i = 1, \ldots, L$$

where I_2 is the identity matrix of order 2 and $T_s = 1[s]$ is the sampling time. The autonomous vehicles are subject to the input constraints

$$\|u^i(t)\|^2 \le 3[m/s^2],\ \forall t \ge 0, \forall i = 1, \ldots, L \tag{12}$$

which are concerned with physical limitations on each vehicle acceleration, and it is required that the distance between two consecutive autonomous vehicles is bounded by $\alpha_{min}^c = 10[m]$ and $\alpha_{max}^c = 25[m]$.

The capabilities provided by the developed co-design have been exploited to build a testing simulation environment, based on real traffic data, for the **DQL**

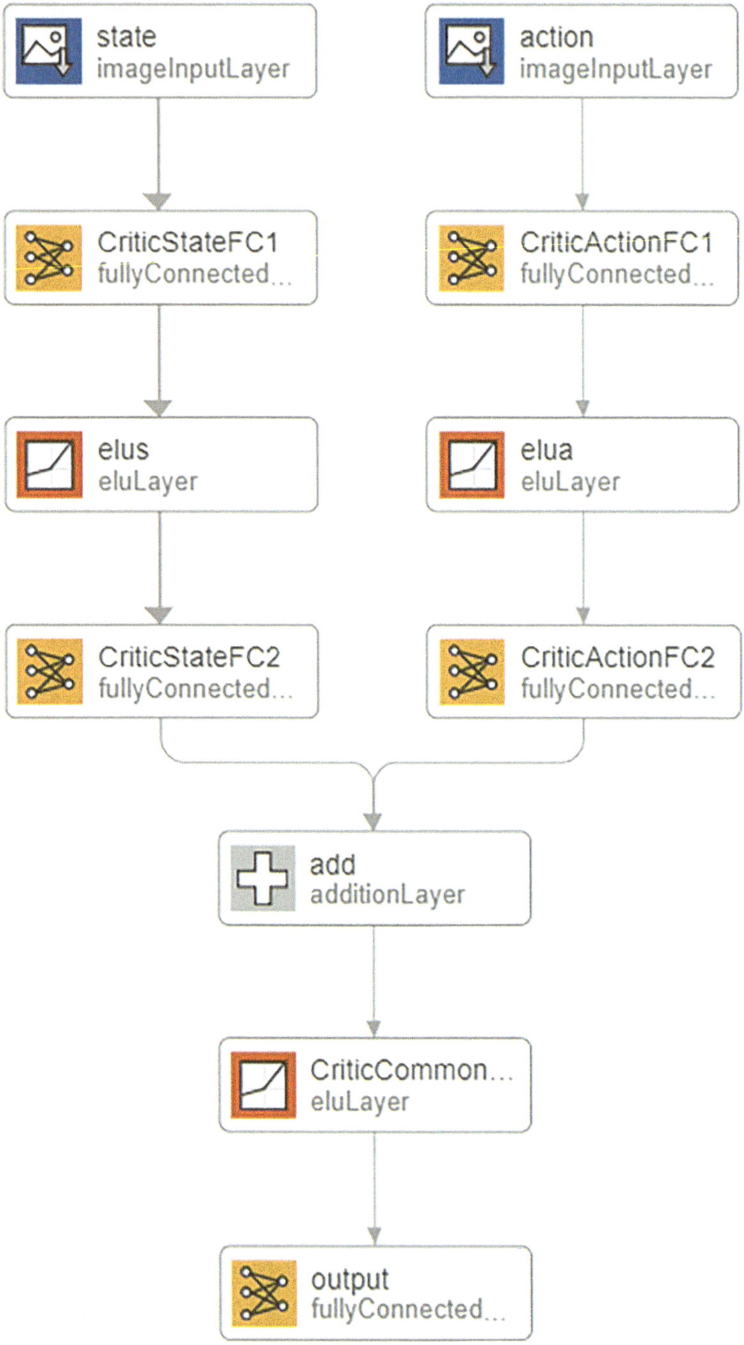

Fig. 7 Neural Network from MATLAB Deep Network Designer

algorithm. The real traffic data set proposed in [15] has been used. These data include representations of the areas around the "Andrea Costa" district. In particular, the traffic data are related to the demand for Bologna's peak hour (8:00 am–9:00 am) and they have been generated considering the higher traffic demand data from March 24, 2010, when a football match took place, and the traffic flow on the day one week in prior. The non-autonomous vehicles in the URN are described in an aggregated manner: the numbers of vehicles to insert are given for certain roads located at the network's border. Following their initial route, these vehicles cross routing decision points at which they get a new route randomly assigned.

The proposed **DQL**-Algorithm is able to deal with actual traffic congestion and to learn from vehicles' experience. More precisely, each autonomous agent benefits from an initial Q-function, pre-trained by means of the built-in routines of the MATLAB Reinforcement Learning Toolbox and of the SUMO data set related to "Andrea Costa" district. This pre-training has been carried out by generating 2000 traffic episodes. Here, each episode is defined as a finite sequence of time instants, with the first instant corresponding to the initial vehicle configuration as shown in Table 1, while the last to the target x^f. Notice that during the training, the NN weights are updated according to (10). Specifically, the training process evolution is depicted in Fig. 8, where the episode reward (the sum of all the rewards received from the agent during each episode), the moving average reward, and the

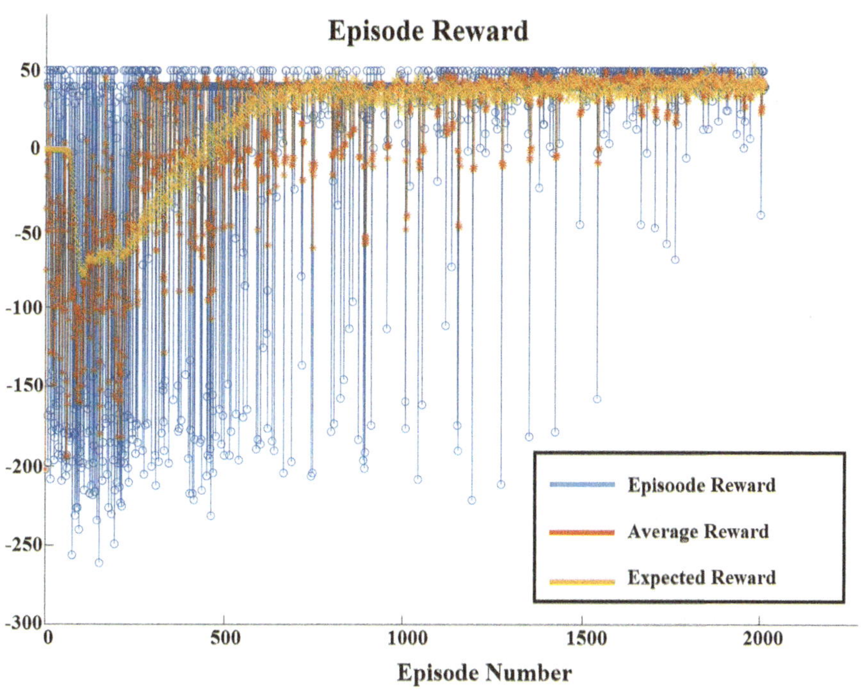

Fig. 8 Training from MATLAB reinforcement learning toolbox

expected episode reward are reported. There it can be observed that the capability of the vehicle to select the more adequate decision improves as the training process evolves: the average reward (red line) increases, while the expected one (orange line) shows an asymptotic convergent behavior to the effective reward. The final part of this plot illustrates the ϵ-greedy policy: even when the agent knows how to obtain the maximum available reward, with a certain small probability, try to explore different routing actions, represented as the blue peaks, at the end of the plot.

7 Conclusions

This work studies the Vehicle Routing Problem in the context of Internet of Vehicles in order to decrease traffic congestion, an issue recently intensified by the rapid growth of vehicles number on the roads. In detail, we propose our Deep Q Learning architecture with a micro-traffic simulations approach. The experimental results show that the proposal has good performance in real scenario of traffic congestion and can effectively reduce the accumulated waiting time. As a consequence, our approach is sustainable because reducing traffic congestion improves environmental pollution.

Future work will consider how to accelerate distributed learning to get a faster vehicle planning decision response and to deal more specifically with the control problem.

Acknowledgments If you want to include acknowledgments of assistance and the like at the end of an individual chapter please use the `acknowledgement` environment—it will automatically be rendered in line with the preferred layout.

References

1. F. Yang, S. Wang, J. Li, Z. Liu, Q. Sun, An overview of internet of vehicles. China Commun. **11**(10), 1–15 (2014)
2. K. Lin, Y. Li, J. Deng, P. Pace, G. Fortino, Clustering-learning-based long-term predictive localization in 5G-envisioned internet of connected vehicles. IEEE Trans. Intell. Transport. Syst. **22**(8), 5232–5246 (2020)
3. D. Cao, X. Wang, L. Li, C. Lv, X. Na, Y. Xing, X. Li, Y. Li, Y. Chen, F.-Y. Wang, Future directions of intelligent vehicles: potentials, possibilities, and perspectives. IEEE Trans. Intell. Veh. **7**(1), 7–10 (2022)
4. M.W. Ulmer, J.C. Goodson, D.C. Mattfeld, B.W. Thomas, On modeling stochastic dynamic vehicle routing problems. EURO J. Transport. Logist. **9**(2), 100008 (2020)
5. P. Toth, D. Vigo, *The Vehicle Routing Problem* (SIAM, Philadelphia, 2002)
6. M. Mangeruga et al., An underwater pathfinding algorithm for optimised planning of survey dives. Remote Sens. **12**(23), 3974 (2020)

7. K. Lin, C. Li, G. Fortino, J. J. Rodrigues, Vehicle route selection based on game evolution in social internet of vehicles. IEEE Internet Things J. **5**(4), 2423–2430 (2018)
8. K. Lin, C. Li, Y. Li, C. Savaglio, G. Fortino, Distributed learning for vehicle routing decision in software defined internet of vehicles. IEEE Trans. Intell. Transport. Syst. **22**(6), 3730–3741 (2020)
9. G. Bono, J.S. Dibangoye, O. Simonin, L. Matignon, F. Pereyron, Solving multi-agent routing problems using deep attention mechanisms. IEEE Trans. Intell. Transport. Syst. **22**(12), 7804–7813 (2020)
10. K. Lin, J. Gao, Y. Li, C. Savaglio, G. Fortino, Multi-granularity collaborative decision with cognitive networking in intelligent transportation systems. IEEE Trans. Intell. Transport. Syst. **24**(1), 1088–1098 (2022)
11. P.A. Lopez, M. Behrisch, L. Bieker-Walz, J. Erdmann, Y.-P. Flötteröd, R. Hilbrich, L. Lücken, J. Rummel, P. Wagner, E. Wießner, Microscopic traffic simulation using sumo, in *2018 21st International Conference on Intelligent Transportation Systems (ITSC)* (IEEE, Piscataway, 2018), pp. 2575–2582
12. F. Giannini, G. Fortino, G. Franzè, F. Pupo, Path planning for vehicle platoons under routing decisions: a distributed approach combining deep reinforcement learning and model predictive control, in *CoDIT'22, Istanbul, Turkey* (2022)
13. F. Giannini, G. Fortino, G. Franzè, F. Pupo, A deep Q learning-model predictive control approach to vehicle routing and control with platoon constraints, in *IEEE 18th International Conference on Automation Science and Engineering (CASE), Mexico City* (2022)
14. F. Giannini, G. Franzè, F. Pupo, G. Fortino, Autonomous vehicles in smart cities: a deep reinforcement learning solution, in *2022 IEEE International Conference on Dependable, Autonomic and Secure Computing, International Conference on Pervasive Intelligence and Computing, International Conference on Cloud and Big Data Computing, International Conference on Cyber Science and Technology Congress (DASC/PiCom/CBDCom/CyberSciTech)* (2022), pp. 1–6
15. L. Bieker, D. Krajzewicz, A. Morra, C. Michelacci, F. Cartolano, Traffic simulation for all: a real world traffic scenario from the city of Bologna, in *Modeling Mobility with Open Data* (Springer, Berlin, 2015), pp. 47–60
16. K.-K. Oh, M.-C. Park, H.-S. Ahn, A survey of multi-agent formation control. Automatica **53**, 424–440 (2015)
17. G. Franzè, W. Lucia, F. Tedesco, A distributed model predictive control scheme for leader–follower multi-agent systems. Int. J. Control **91**(2), 369–382 (2018)
18. R.S. Sutton, A.G. Barto, *Reinforcement Learning: An Introduction* (MIT Press, Cambridge, 2018)
19. J. Fan, Z. Wang, Y. Xie, Z. Yang, A theoretical analysis of deep Q-learning, in *Learning for Dynamics and Control* (PMLR, 2020), pp. 486–489
20. P.A. Lopez, M. Behrisch, L. Bieker-Walz, J. Erdmann, Y.-P. Flötteröd, R. Hilbrich, L. Lücken, J. Rummel, P. Wagner, E. Wießner, Microscopic traffic simulation using sumo, in *The 21st IEEE International Conference on Intelligent Transportation Systems* (IEEE, Piscataway, 2018) [Online]. Available: https://elib.dlr.de/124092/
21. B.-L. Ye, W. Wu, K. Ruan, L. Li, T. Chen, H. Gao, Y. Chen, A survey of model predictive control methods for traffic signal control. IEEE/CAA J. Autom. Sin. **6**(3), 623–640 (2019)
22. S. Chen, Z. Wu, D. Rincon, P.D. Christofides, Machine learning-based distributed model predictive control of nonlinear processes. AIChE J. **66**(11), e17013 (2020)
23. M.M. Alam, T. Ahmed, M. Hossain, F. Pupo, G. Fortino, Federated ensemble-learning for transport mode detection in vehicular edge network. Future Gener. Comp. Sys. **149**, 89–104 (2023)
24. A.F. Acosta, J.E. Espinosa, J. Espinosa, Traci4matlab: enabling the integration of the sumo road traffic simulator and Matlab® through a software re-engineering process, in *Modeling Mobility with Open Data* (Springer, Berlin, 2015), pp. 155–170

25. Reinforcement learning toolbox: User's guide (r2022a). [Online]. Available: www.mathworks.com/help/reinforcement-learning/
26. Deep learning toolbox: User's guide (r2022a). [Online]. Available: www.mathworks.com/help/deeplearning/
27. G. Henderson, Applied mathematics in integrated navigation systems–third edition RM rogers American institute of aeronautics and astronautics, 1801 Alexander Bell Drive, Suite 500, Reston, VA 20191–4344, USA. 2007. 408pp. illustrated. 94.95 (non-members). ISBN 1-56347-927-3. Aeronaut. J. **113**(1141), 202–202 (2009)

Compressed Sensing-Based IoMT Applications

Bharat Lal, Qimeng Li, Raffaele Gravina, and Pasquale Corsonello

1 Introduction

The Internet of Medical Things (IoMT) has revolutionized the healthcare industry by enabling the development of wearable sensors, remote monitoring systems, and telemedicine devices that can monitor patients' health and transmit data in real time. However, the growing volume of healthcare data generated by these devices presents significant data storage, processing, and transmission challenges. Compressed sensing (CS) is a signal processing technique that can help to address these challenges by enabling the efficient acquisition and compression of signals. CS is beneficial for Internet of Things (IoT)-based healthcare devices because it can help to reduce the amount of data that needs to be transmitted, stored, and processed, improving the healthcare system's efficiency.

This chapter provides an overview of CS approaches for IoMT devices. It begins by introducing the basic principles of CS and its applications in healthcare. It then describes the challenges and opportunities associated with CS in healthcare, including selecting measurement matrices, signal reconstructions, and compressed learning techniques. The chapter also discusses the various CS algorithms that can be used in IoMT, such as optimization algorithms, greedy algorithms, and Bayesian methods. It provides an overview of the strengths and limitations of these algorithms and discusses how they can be optimized for healthcare applications. Finally, the chapter includes case studies demonstrating CS's effectiveness in healthcare, such as monitoring electroencephalography (EEG), electrocardiography (ECG), electromyography (EMG), and electrodermal activity (EDA) signals. These case

B. Lal (✉) · Q. Li · R. Gravina · P. Corsonello
Department of Informatics, Modeling, Electronics, and Systems, University of Calabria, Arcavacata, Italy
e-mail: bharat.lal@dimes.unical.it; qimeng.li@unical.it; r.gravina@dimes.unical.it; pasquale.corsonello@unical.it

C. Savaglio et al. (eds.), *Device-Edge-Cloud Continuum*, Internet of Things,
https://doi.org/10.1007/978-3-031-42194-5_11

studies cover a range of applications, including wearable sensors, remote monitoring systems, and telemedicine devices. The case studies illustrate how CS can improve healthcare data acquisition, processing, and transmission accuracy and efficiency.

2 Overview of IoMT

The IoMT refers to the integration of medical devices and sensors with the Internet. IoMT enables healthcare providers to collect and analyze real-time patient data, monitor patients remotely, and improve patient outcomes. The IoMT is rapidly advancing, and it has the potential to transform the healthcare industry. Medical devices and equipment can now be connected to the Internet, allowing healthcare providers to access critical patient data and make more informed decisions about treatment. This technology can improve patient outcomes and reduce healthcare costs by providing more personalized and efficient care. Applications of IoMT include wearable devices that monitor the signals such as ECG, EEG, EMG, and GSR remotely and allow doctors to monitor patients outside of the hospital setting. The data generated by these devices can be analyzed using artificial intelligence and machine learning algorithms, providing healthcare providers with insights into patient health and potential treatments. However, the IoMT also presents challenges related to big data generated from sensors, computational resources, data privacy, security, and interoperability. Healthcare providers must ensure that patient's data is secure and protected from unauthorized access and that devices and systems can communicate with each other seamlessly. As the IoMT continues to develop, it will be important to address these challenges to ensure that the benefits of this technology can be realized.

2.1 IoMT Framework

According to the ITU (International Telecommunication Union), the IoT is a worldwide infrastructure that connects physical and virtual objects through information and communication technologies to provide advanced services. A smart healthcare system based on IoMT consists of various smart medical devices connected through the Internet [1]. The IoMT framework for smart healthcare involves multiple stages. Initially, smart sensors were integrated to collect medical data from the patient's body. These devices are connected through either a body sensor network (BSN) or a wireless sensor network (WSN) [2]. Next, the collected medical data is transmitted over the Internet to the next phase, which deals with prediction and analysis. Proper artificial intelligence (AI)-based data transformation and interpretation techniques can be utilized for analysis [3]. In critical situations, medical personnel can be alerted using AI-based applications on smartphones [4, 5]. For less severe cases, preventive measures can be taken by the patients themselves.

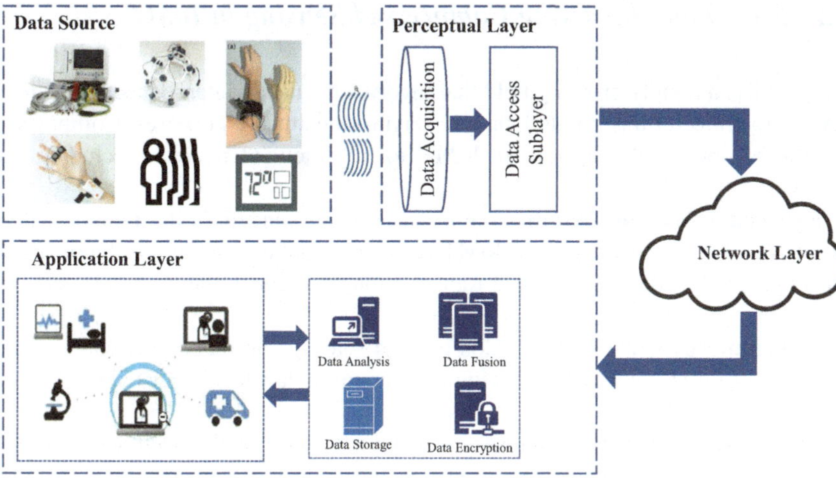

Fig. 1 IoMT architecture

Over time, numerous structures have been suggested by researchers to create a standardized structure for IoT applications in healthcare. These structures take into account various factors associated with IoT gadgets, communication protocols, and networking, as well as the intended applications and services. For instance, Sun et al. [6] explained that the IoMT architecture consist of three layers as shown in Fig. 1: the application layer, perceptual layer, and network layer. The perceptual layer, which is the bottom layer, collects data from the source and analyzes it. It consists of two sublayers: data access and data acquisition. The data acquisition sublayer processes the collected data using various medical perception equipment and signal acquisition equipment such as graphic code, RFID, and GPRS. The data access sublayer connects the collected data to the network layer using short-range data transfer techniques like Bluetooth, Wi-Fi, and Zigbee.

The network layer is responsible for offering platform and interface-related services and employs different data transmission methods. It comprises two layers: the service layer and the network transmission layer. The network transmission sublayer employs technologies such as mobile communication networks, wireless sensor networks (WSNs), and the Internet to transmit data received from the perception layer accurately, consistently, and in real time. The service layer integrates various networks, data warehouses, and information description formats, and it provides open interface services and other platform-related services to achieve such integrations.

The application layer uses network layer data to handle medical records through different applications. It has two sublayers: medical information applications and medical decision-making applications. The latter holds health equipment and materials for inpatient/outpatient info and treatment records. The former analyzes data on patients, diseases, medications, diagnoses, and treatments.

2.2 *Motivation for Using Compressed Sensing in IoMT*

The IoMT is a rapidly growing field that involves the use of interconnected medical devices to collect, transmit, and analyze data for healthcare purposes. Compressed sensing can be a valuable tool in IoMT for a variety of reasons, including the following:

Reduced data transmission: Compressed sensing can reduce the amount of data that needs to be transmitted from IoMT devices to a central server or cloud, which can improve the efficiency of data transmission and reduce the strain on network resources.

Lower-energy consumption: Compressed sensing can reduce the energy consumption of IoMT devices by reducing the amount of data that needs to be collected and transmitted.

Improved accuracy: Compressed sensing can improve the accuracy of IoMT devices by reducing noise and artifacts in the collected data, resulting in more reliable and precise readings.

Improved privacy and security: Compressed sensing can protect sensitive data by using random sensing matrices that needs to be transmitted from IoMT devices, which can improve privacy and security for patients and healthcare providers.

Real-time monitoring: Compressed sensing can enable real-time monitoring of patients, allowing healthcare providers to quickly identify and respond to potential health issues.

In general, compressed sensing can help to improve the efficiency, accuracy, and security of IoMT devices, which can lead to better healthcare outcomes and a more seamless patient experience.

3 Objectives and Scope

The CS framework for IoMT has found widespread use in acquiring physiological signals such as EEG, ECG, EMG, and GSR, compressing and reconstructing them, extracting their features, and classifying them. This approach has been particularly useful in addressing challenges such as faster signal acquisition, quality reconstruction, and energy efficiency. The advancements in the acquisition and processing of physiological signals have greatly improved the understanding of disease detection and diagnosis in patients. However, with the increasing data volume and power consumption, it has become crucial to employ effective techniques for data compression, signal acquisition, transmission, and processing, particularly in telemonitoring healthcare applications.

A new area of research involves combining compressed sensing (CS) with physiological signals to overcome the challenges posed by the large amount of data generated from such signals, limited transmission bandwidth, and power consumption. CS has the potential to improve the accuracy of medical diagnosis and prediction while reducing data storage and transmission requirements. The

CS approach involves simultaneously acquiring and compressing the physiological signals, followed by optimization or machine learning algorithms to extract relevant features and perform tasks such as classification, clustering, or prediction. This approach can be used in remote patient monitoring, where compressed physiological signals can be transmitted over low-bandwidth networks and analyzed in real time to provide feedback to healthcare providers.

Therefore, this research aims to identify the CS techniques for IoMT and their applications, covering various aspects such as acquisition strategies, reconstruction algorithms, measurement matrices, sparse bases, and evaluation metrics, with a specific focus on reconstruction-free CS techniques and potential hardware implementations.

4 Compressed Sensing Overview

The compressed sensing theory, also known as compressive sensing, was initially proposed by Candès and Tao [7]. This approach has proven to be an effective method for acquiring and reconstructing signals by taking advantage of the fact that physical signals are typically sparse in their original form or on a well-selected basis (e.g., Fourier or wavelet transform). By directly acquiring a compressed form of the signal in a single stage, CS employs mathematical tools like the $L1$ norm to represent or reconstruct the original sparse signal from only a few samples at the receiving node. CS has gained widespread attention from researchers in recent years in various fields, including compressive imaging [8–10], biomedical applications [11, 12], communication systems [13], pattern recognition [13], and speech processing [14]. The remarkable advantage of CS is that it can better reconstruct the original signal using a small number of linear projections.

Traditional sensing and processing theories rely on the Nyquist-Shannon sampling theorem, which states that to preserve the information of a signal with a bandwidth of W, it should be sampled at a secured sampling frequency $fs \geq 2W$. However, with the increasing number of digital sensing devices, challenges have arisen, such as the large amount of data generated, high data transmission rates, and constraints on sampling/processing platforms. To address these challenges, compression techniques are often employed by discarding redundant samples in the signal, leading to shorter communication delays, efficient bandwidth usage, and reduced battery drain.

In WSN applications, energy consumption in the sensing nodes is mainly dominated by wireless communication rather than onboard processing, so data reduction strategies at the sensor node are necessary to minimize the energy cost of the system. The recently proposed CS theory suggests that for a specific type of structured signal, it is possible to reduce the sampling rate without significant loss of data information. If the signal is sparse or compressible, it is possible to capture all the information without any loss by acquiring fewer samples than what the Nyquist-Shannon theorem states.

CS differs significantly from state-of-the-art compression techniques as it is a one-stage framework that performs both acquisition and compression simultaneously, acquiring the data in a compressed form at the sensor and efficiently reconstructing the original data from a fewer sample signal at the receiver. CS leverages the fact that most real-world signals exhibit sparse behavior when represented by an appropriate basis, and rather than sampling the signal at a Nyquist rate and then performing compression, CS takes random measures of the signal in a sub-Nyquist rate corresponding to the signal information rate.

4.1 Compressed Sensing for Physiological Signals

Physiological signals such as EEG, ECG, and EMG are often sparse or compressible in some domain or basis. CS can be used to efficiently acquire and process these signals using a small number of measurements. CS for physiological signal acquisition and reconstruction block diagram is shown in Fig. 2. The compressed signal needs to be acquired first using different measurement matrices, such as random, structured, and deterministic. Then, it needs to be transmitted to a remote location and reconstructed using multiple reconstruction algorithms.

The mathematical formulation of compressed sensing for physiological signals is as follows: Let x be an $N \times 1$ vector representing the original physiological signal, which is assumed to be sparse or compressible in some basis ψ. Let y be an M x 1 vector representing the linear measurements of the signal, which are obtained by multiplying the signal x by a sensing matrix A of size $M \times N$ as shown in Eq. 1:

$$y = Ax \tag{1}$$

Physiological signals like EEG, ECG, and EMG exhibit sparsity and high compressibility. This means that by using an appropriate basis to represent the signal, only a few modes are active, resulting in a smaller number of values needed for an accurate representation. An alternative approach is to transform the compressible signal $x \in \mathbb{R}^n$ into a sparse matrix $s \in \mathbb{R}^n$ using a transform basis, such as the Fourier or wavelet basis $\psi \in \mathbb{R}^{n \times n}$. If the vector s contains only K nonzero elements, it is referred to as K-sparse in ψ:

$$x = \psi s = \sum_{i=1}^{N} \psi_i s_i \tag{2}$$

Further, it can be written as

$$y = A\psi s \tag{3}$$

where

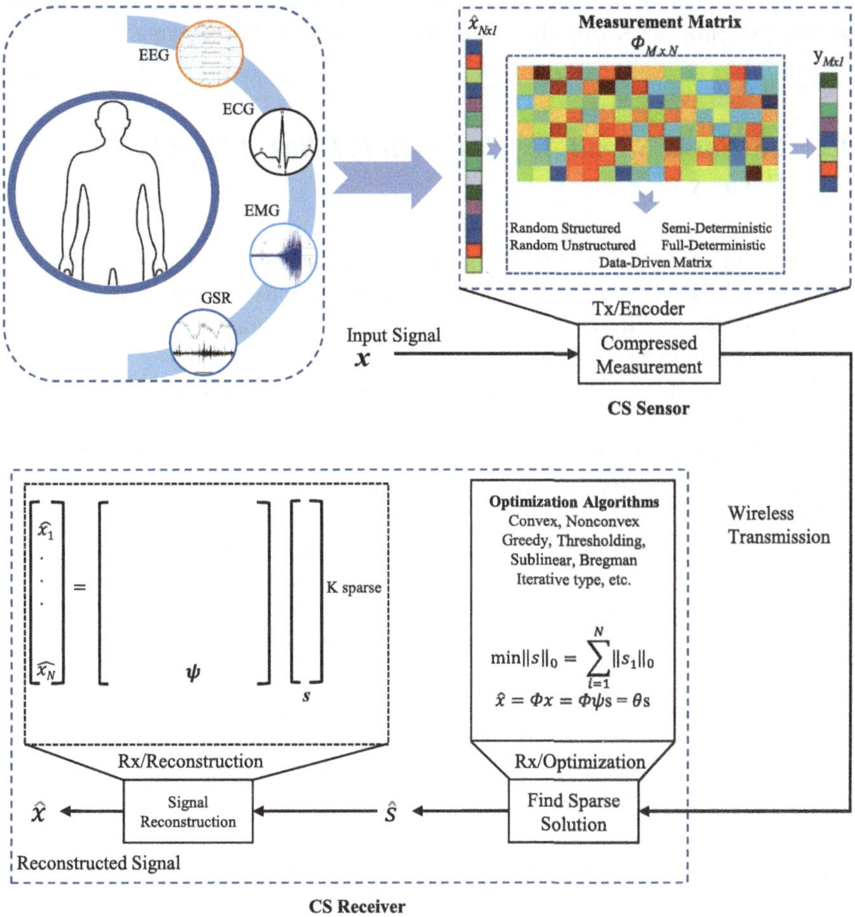

Fig. 2 CS-based signal acquisition and reconstructions

$$\theta = A\psi \tag{4}$$

In the case of physiological signals, the sensing matrix A is typically a random or pseudorandom matrix, such as a Gaussian random matrix or a Bernoulli random matrix. These matrices have been shown to be effective in capturing the important information in physiological signals while minimizing the number of measurements required. Once the compressed measurements y have been obtained, they can be used to reconstruct the original signal x using the optimization problem. The reconstructed signal $\hat{x}$ will be sparse or compressible in the same basis as the original signal x, allowing for efficient processing and analysis. CS has many applications in the field of physiological signal processing, including the compression and denoising of ECG and EEG signals, as well as the localization of brain activity from

EEG signals. It allows for the efficient acquisition and processing of these signals, leading to significant savings in storage, transmission, and processing.

4.2 *CS-Based Signal Acquisition in IoT-Based Healthcare Devices*

In IoT healthcare devices, CS can be used for compressed data acquisition from various devices, such as ECG, EEG, EMG, GSR sensors, magnetic resonance imaging (MRI), etc. To implement the CS-based data acquisition in IoT healthcare devices, the following steps can be taken:

Signal model: The signal of interest, such as EEG, ECG, EMG, and GSR, should be modeled in a way that allows for sparse representation or compressibility. This can be achieved using techniques such as wavelet transforms, discrete cosine transforms, data-driven dictionary or other signal processing methods.

Measurement matrix design: A measurement matrix is used to compress the signal by acquiring a smaller number of measurements than traditional sampling methods, i.e., Nyquist sampling. The matrix should be designed to satisfy the restricted isometry property (RIP), which ensures that the compressed measurements contain enough information to reconstruct the original signal. Furthermore, sufficient conditions for perfect signal recovery are defined in Sect. 4.3.

Signal reconstruction: The compressed measurements can be used to reconstruct the original signal using techniques such as convex optimization, sparse coding, or iterative algorithms. Moreover, reconstruction algorithms are defined in Sect. 4.2.2.

Compressed learning: In some applications, signals do not need to be reconstructed at the receiving end. Therefore, important features can be extracted directly from compressed measurements. Furthermore, this concept is explained in Sect. 4.4.

Validation: The reconstructed signal should be validated to ensure that it is accurate and can be used for further analysis.

4.2.1 Measurement Matrices

To measure a signal accurately, a measurement matrix is used to sample representative elements of the signal. The accuracy and processing time of the measurement process depend heavily on the number of measurements, which varies with each matrix. Thus, it is crucial to design precise measurement matrices. Various measurement matrices have been proposed in the literature over the last ten years [15] as shown in Fig. 3. Random matrices, which are commonly used, have the disadvantage of requiring more measurements and cannot be stored or regenerated at the receiving end. Structured matrices are faster to acquire, require less storage space, and reduce transmission overheads, but they sacrifice universality. Deterministic matrices, which meet RIP, have little mutual coherence.

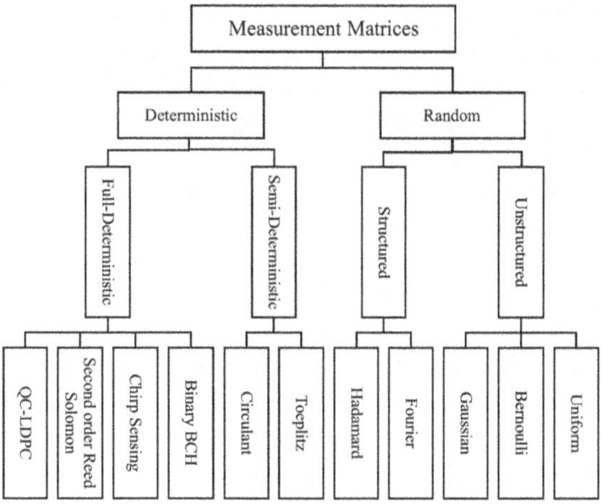

Fig. 3 Measurement matrices

Semideterministic matrices allow for quick decoding and can be applied to several applications, while pure deterministic matrices are based on mutual coherence and RIP. It has been suggested in the literature that deterministic matrices are a good option for practical applications, such as physiological signal acquisitions [16, 17]. For deterministic matrices, however, it is necessary to make a greater number of measurements in order to reduce the reconstruction error. Therefore, when some preliminary information about the location of nonzero elements of the sparse signal is available, the researcher must design a specific structure to reduce sampling rate, recovery error, processing time, and matrix size flexibility. Developing explicit measurement matrices remains a challenge for researchers.

The measurement of physiological signals plays a crucial role in medical monitoring. To effectively monitor patients remotely, wireless sensors are used, but it is important to ensure that they consume low energy for continuous monitoring. CS-based sampling is a promising solution to ensure low-power consumption for continuous patient monitoring. This method transfers the computational burden to the receiver where signal reconstruction can be done offline. Researchers have used various CS acquisition methods to reduce power consumption (PC) in wireless healthcare applications. For instance, Mamaghanian et al. [18] developed a CS-based sampling node for ECG signals, extending battery life by 37.1% compared to its Nyquist counterpart. CS has also proven successful in acquiring EEG [19, 20], ECG [21, 22], EMG [23], GSR [24] signals, and several CS acquisition techniques have already been implemented on hardware [25, 26].

4.2.2 Signal Reconstructions

The reconstruction matrix θ and compressed measurement vector y are applied to the reconstruction algorithm to find the sparse solution as illustrated in Fig. 2.

Equation 3 is an under-determined system of linear equation and have multiple possible solutions; the optimization algorithms are used to find the sparse solutions. Once we find the sparse solution, then it is possible to retrieve the original signal from the sparse vector by applying the inverse transform. Given the measurement vector y, there exists a sparse vector s that is consistent with it. The primary goal of optimization algorithms is to find the vector $\hat{s}$ that has the fewest nonzero elements and satisfies the given conditions:

$$\hat{s} = \min \|s\|_0 \quad \text{subject to } y = A\psi s \tag{5}$$

The l_o-pseudonorm, denoted by min $\|.\|_0$, counts the number of nonzero entries in a vector s [27]. The problem of finding the solution is non-convex, and searching for all possible solutions is computationally expensive, even for standard-sized problems. Therefore, it is classified as an NP-hard problem. However, alternative methods have been proposed in the literature which relax the optimization problem by replacing it with a convex l_1-minimization problem, as shown in Eq. 6, under certain conditions on the A measurement matrix. Solving l_1-minimization problems can be achieved in near polynomial time using linear programming solvers, and the l_1-norm approach promotes sparsity, whereas the l_2-norm approach does not:

$$\hat{s} = \min ||s||_1 \quad \text{subject to} y = C\psi s \tag{6}$$

The algorithm for CS reconstruction produces a sparse representation of the desired signal x, denoted by $\hat{s}$, and the recovered signal $\hat{x}$ can be obtained from $\hat{s}$ by applying the inverse transform. CS reconstruction algorithms are divided into many categories, and few of them are given in Fig. 4.

Each method has its own advantages and disadvantages. Convex optimization is capable of reconstructing the original signal from a smaller set of measurements, but it has a higher computational complexity [28, 29]. Greedy algorithms, on the other hand, are typically fast and have lower time complexity, but they require a matrix-inverse operation in each iteration, which can be computationally expensive and require more expensive hardware [30, 31]. Combinatorial algorithms can rapidly reconstruct data, but they require a particular type of structured samples that may be difficult to obtain in practical applications [32, 33]. Non-convex algorithms, despite being difficult to implement, require fewer measurements to recover a signal, perform better even with weaker RIP for larger signals, and have a complexity similar to that of convex optimization technique [34]. The Bregman algorithm provides a sparse output and is fast, but it has very high computational costs [35, 36]. The deep compressed learning techniques consistently demonstrate faster processing times, often faster than traditional reconstruction algorithms. This makes them suitable for real-time computing applications that are widely used [37].

CS Algorithm	Subcategories
Convex Optimization Complexity: ≈ $0(M^2N^3)$	BB-Simplex BP-Interior Fixed Point Continuation Gradient Projection Total Variation denoising Basis Pursuit (BP) Basis Pursuit denoising Dantzig Selector
Combinatorial/ Sublinear Algorithms Complaxity: ≈ N/A	Chaining Pursuits Heavy Hitters on Steroids (HSS) Fourier Sampling Algorithm
Non-Convex Complexity: ≈ $0(M^2N^3)$	Focal Understanding System Solution Iterative Re-weighted Least Squares Sparse Bayesian Learning Algorithm Monte-Carlo based algorithm
Bregman Iterative type Complexity: ≈ $0(NM^2)$	Linearized Bregman Logistic Bregman Split Bregman
Thresholding Complexity:≈ $0(NM\ \#iteration)$	Iterative Soft Thresholding (IST) Iterative Hard Thresholding (IHT) Message Passing Algorithms Expander Matching Pursuit (EMP) Sparse Matching Pursuit (SMP) Sequential SMP Belief Propagation
Greedy Complexity: ≈ $0(NM\ \#iteration)$	Matching Pursuit (MP) Gradient Pursuit (GP) Orthogonal MP (OMP) Regularized OMP (ROMP) Compressive Sampling MP (Cosamp) Subspace Pursuit (SP)
Deep learning-based algorithms	Deep Compressive Sensing, Deep Decoder and Learned Iterative Shrinkage Thresholding

Fig. 4 Reconstruction algorithms

4.3 Compressed Learning

Compressed learning (CL) is a machine learning technique that focuses on extracting features from direct compressed measurement and reducing the amount of data required for training models. This approach is particularly useful in IoT healthcare applications, where signal reconstruction can be expensive and time-consuming. By using compressed learning, we can save the computational resources at receiving end, and important features can be reconstructed from direct compressed measurements. One way to apply compressed learning in IoT healthcare applications is by using ML algorithms that can learn from sparse data. CL can help to reduce the amount of data required for training models in IoT healthcare applications. By

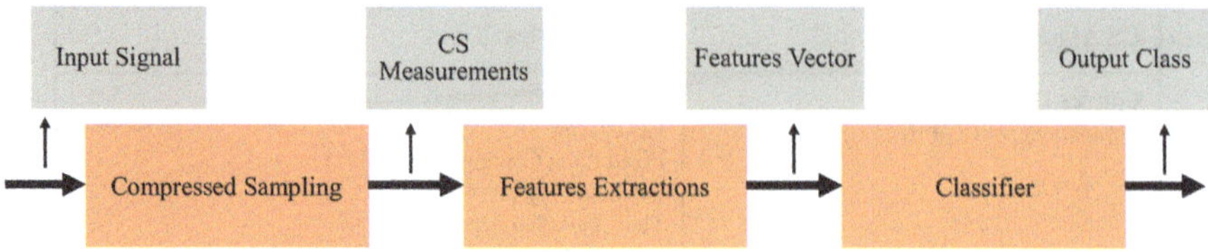

Fig. 5 Compressed learning approach

optimizing the use of data, we can improve the accuracy of our models and reduce the costs associated with data collection and storage. Furthermore, the compressed learning block diagram is shown in Fig. 5.

Various studies [38–40] have been conducted to analyze physiological signals directly on CS measurements without the need for time-consuming signal reconstruction. Essentially, CS reconstruction involves learning the signal from linear CS measurements using iterative algorithms, and integrating this learning with ML can be a clever approach. However, there has not yet been a systematic approach to compressed learning for physiological signals, although some studies have implemented compressed learning. Ching-Yao Chou et al. [41] proposed a biometric user identification method using ECG, known as the compressed alignment-aided compressive analysis (CA-CA) algorithm. This algorithm uses a PCA-based dictionary in the compressed domain to directly recover information from compressed data without reconstructing the ECG signal. As a result, the computation time is reduced by 81.08%, and the compressed ECG signal is classified with high accuracy. Compared to compressed learning (CL), the proposed algorithm improves accuracy by 7.11%. However, the accuracy of the algorithm drops compared to reconstruction learning with alignment (RL-A), and it has only been tested on a small ECG database.

4.4 CS Conditions for Exact Recovery

In order to ensure the reliability and precision of the restored signals obtained from compressed measurements, certain criteria must be met by the CS parameters, including the sensing matrix and sparse bases.

4.4.1 Restricted Isometry Property

Compressed sensing techniques aim to reconstruct signals using a small number of measurements. The restricted isometry property (RIP) is a mathematical concept used to evaluate the performance of these techniques. If the vector s is k-sparse in the transform domain, then the matrix θ must satisfy the RIP of order k in order to recover s from compressed measurement y. This means that for any k-sparse vector u with the same nonzero entries as s, the following inequality must hold:

$$1 - \delta \leq \frac{|\theta u|2}{|u|2} \leq 1 + \delta \tag{7}$$

where δ is a positive constant that satisfies the RIP, in which the inequality ensures that the matrix θ preserves the distance between any two k-sparse vectors. In order to obtain a robust solution, the matrix θ must satisfy Eq. 7. However, calculating δ can be challenging. Additionally, coherence is another condition that guarantees a stable solution.

Essentially, the RIP ensures that A preserves the inner product and norm of any k-sparse vector with minimal error. Satisfying the RIP allows for the recovery of any k-sparse vector x from its compressed measurements Ax with high probability, using convex optimization techniques. The RIP is an important property in compressed sensing as it guarantees the success of sparse recovery algorithms under certain conditions. Many natural classes of matrices, including random Gaussian matrices and certain structured matrices, have been shown to satisfy the RIP with high probability.

4.4.2 Incoherency

To ensure an accurate reconstruction, the measurement matrix A and the sparse basis ψ must be incoherent with each other, as stated by the condition. Equation 8 explains the relationship between the coherence of the two matrices. The coherence value $\mu(A, \Psi)$ is calculated by considering the highest correlation between any two elements of the selected pair of matrices, with a range of $\mu(A, \Psi) \in \left[1, \sqrt{n}\right]$. The number of measurements required for CS reconstruction varies depending on the coherence value, with lower coherence values requiring fewer measurements. This is because Eq. 7 reveals that the coherence value is the maximum absolute inner product between the rows of A and the columns of ψ, and when this value is smaller, fewer measurements are required for accurate reconstruction:

$$\mu(A, \Psi) = \sqrt{n} \max_{1 \leq ij \leq n} |\langle A_i, \psi_j \rangle|. \tag{8}$$

5 Compressed Sensing-Based IoMT

CS-based IoMT is increasingly important in healthcare for medical diagnosis and treatment. General CS-based IoMT architecture is shown in Fig. 6. The sensing layer of an IoT platform in healthcare comprises smart devices that collect various health-related data types. These devices are deployed in different locations depending on the targeted healthcare application and gather data over an extended period.

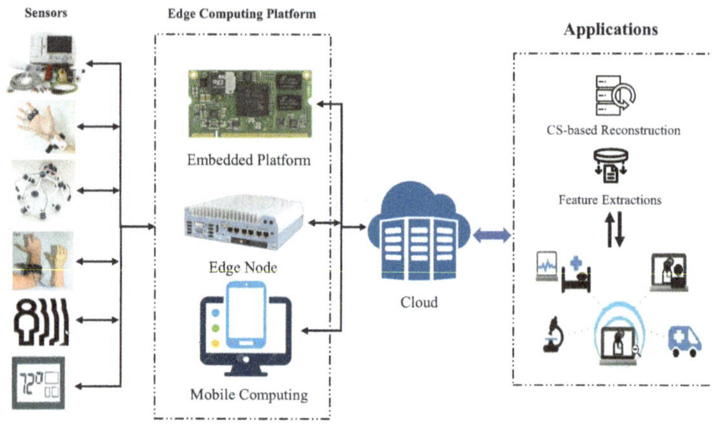

Fig. 6 CS-based IoT healthcare

Power consumption is a critical issue in healthcare IoT, and CS can minimize power consumption during data acquisition and transmission by acting as a compression technique. The local processing unit stores and aggregates the compressed data transmitted from the sensors, and embedded systems on the local processing unit can be designed as a "CS-based edge computing platform." This platform can extract salient information and features from the compressed data and send it to the cloud, which hosts the application layer. In healthcare, CS can be used with other data analytic algorithms to execute specific tasks, providing solutions for medical diagnosis and treatment. Deploying CS at all three levels of the IoT platform in healthcare can achieve significant reductions in power consumption and network bandwidth while improving the efficiency and accuracy of healthcare data analysis.

5.1 Sensing Layer

The sensing layer of CS-based IoMT (Internet of Medical Things) refers to the layer responsible for capturing data from medical devices and sensors. It is a crucial component of the overall IoMT architecture, as it forms the foundation for the collection of data that will be used to provide insights and support medical decision-making. The sensing layer consists of a variety of medical devices and sensors that are designed to capture data on various aspects of patient health, such as heart rate, blood pressure, blood glucose levels, temperature, EEG, and EMG. These devices and sensors can be worn by patients, implanted in their bodies, or placed in clinical settings. However, the major challenge in the sensing layer is the high-power consumption of radio frequency transmission, which can decrease the performance of the system. Various efforts have been made to develop low-power, low-cost compression techniques; however, the CS-based technique consumes less power and increases the battery life. To minimize power consumption, the best sensing matrix can be applied such as adaptive measurements and deterministic measurement technique. The data captured by the sensing layer is typically transmitted to a centralized system, where it is analyzed and processed using various algorithms and machine learning techniques. The insights generated from this data can be used to provide real-time monitoring of patients, support clinical decision-making, and even predict potential health issues before they occur.

5.1.1 Adaptive Measurements

Due to the complexity and non-stationarity of physiological signals, compressed sensing techniques that utilize adaptive measurements have been proposed to improve their reconstruction accuracy. Adaptive compressed sensing in physiological signal processing involves selecting the most informative measurements dynamically based on the signal characteristics. This can be achieved by using a sensing matrix that adapts to the signal, such as random matrices or orthogonal matching pursuit. In addition, signal-specific properties, such as sparsity or correlation between samples, can be used to adapt the sensing matrix to improve the signal reconstruction quality. One approach for adaptive compressed sensing in physiological signals is to use wavelet transform-based sparsity. Wavelets are a popular tool for signal decomposition, and they can be used to represent signals in a sparse domain. By exploiting the sparsity in the wavelet domain, it is possible to achieve high-quality signal reconstruction with fewer measurements. Another approach is to use machine learning algorithms such as deep neural networks to adaptively learn the sensing matrix based on the signal characteristics. These methods can improve the accuracy of compressed sensing by exploiting the correlations between different physiological signals or by utilizing prior knowledge about the signal structure. CS with adaptive measurements has shown promise in improving the reconstruction accuracy of physiological signals while reducing the number of measurements required for signal acquisition. It has the potential to enable more efficient and effective signal processing for a range of physiological applications.

5.2 Processing Layer

Wireless sensor networks (WSNs) are widely used in the IoT platforms to generate big data. Although big data provides comprehensive knowledge about the sensed phenomena, it requires powerful storage, processing, and information-retrieving mechanisms that commonly used computing environments cannot handle. This limitation reduces the performance of IoT platforms by increasing high-level service requirements. To overcome this challenge, gateway/fog computing is proposed as a trade-off solution. It involves installing IoT gateway devices on powerful computing embedded platforms, such as ARM Cortex-A, to bring the computing platform to the edge of the network. This approach is known as edge computing, and its goal is to analyze time-sensitive data at a platform in proximity to where the data were initially collected, rather than sending all data to the cloud.

FPGA (field-programmable gate array) technology is a promising environment to implement edge computing platforms due to its energy efficiency, interactivity, powerful computation abilities, and reconfigurable embedded systems. The hardware implementation of data reconstruction algorithms for edge computing platforms is a suitable approach. It provides an effective solution to reconstruct the original data from compressed data generated at the edge nodes of the network. This approach allows for a more efficient use of the limited resources available in edge computing devices, making them capable of handling complex tasks. The use of edge computing platforms based on FPGA technology is an effective solution to the performance limitations of IoT platforms caused by big data. The hardware implementation of data reconstruction algorithms is a promising approach to enable these platforms to handle complex tasks and achieve high-level service requirements.

5.3 *Application Layer*

The IoMT is a network of medical devices and applications that are connected to healthcare IT systems through the Internet. The application layer, which is responsible for managing the data and functionality of the medical devices and applications. The application layer of the IoMT is responsible for providing the user interface for the medical devices and applications, as well as for managing the data flow between them. This layer provides a high-level view of the data generated by the medical devices and applications and enables healthcare providers to access and analyze this data in real time. The application layer of the IoMT also includes various software and platforms that enable healthcare providers to analyze and visualize the data generated by the medical devices and applications. These tools enable healthcare providers to make informed decisions about patient care and treatment, and also help to improve patient outcomes.

6 Challenges and Future Directions

The CS technique has shown promise in addressing data acquisition and transmission challenges in the IoMT. Energy-efficient CS encoders and structured sensing matrices in the sensing layer can decrease system complexity and improve energy efficiency. For the processing layer, using gateway embedded on many-core platforms instead of single-core platforms can improve speed and quality to meet real-time requirements. Additionally, incorporating CS-based encryption methods can enhance security and privacy for specific healthcare IoT applications. However, there are still several challenges and trends that need to be considered in the application of CS-based IoMT.

One significant challenge is the trade-off between the compression ratio and signal quality. A higher compression ratio often leads to a lower signal quality, which can affect diagnosis and treatment decisions. Thus, there is a need to strike a balance between the two to ensure optimal performance.

The choice of reconstruction algorithm depends on the parameters, and for accurate reconstruction where computation time is not critical, BSBL and BPDN are the recommended choices. Threshold-based reconstruction is suitable for real-time applications where computation time is not the primary concern. However, BSBL is not suitable for real-time multichannel signals as it recovers signals channel by channel. Most CS algorithms are not energy efficient for wireless telemonitoring, so the compression process should be kept simple. It is also important to note that all signals generated by medical sensors are not sparse, so CS techniques may not be applicable to all signals. Multiple algorithms can be used for signal reconstruction, but this increases computational complexity. To reduce complexity, reconstruction-free CS is recommended as certain features can be extracted directly from compressed measurements without the need for signal reconstruction.

Another challenge is the need to consider the privacy and security of patient data. With the increased use of IoMT, patient data are becoming more vulnerable to attacks from malicious attackers. Therefore, it is essential to develop secure and privacy-preserving algorithms that can protect patient data from unauthorized access.

CS techniques offer power and latency savings for wearable devices, as they directly extract compressed information at the time of acquisition, reducing memory storage and improving speed and energy efficiency. Articles [42, 43] proposed CS-based methods to save power and reduce latency in IoT applications, including WSN and deep CS networks. In healthcare, there is a need for a personalized data-driven approach to treatment, which requires a quantitative integrative tool to translate clinical observations to the individual level. While CS is useful for acquiring and reconstructing physiological signals, reconstruction-free CS can be more beneficial for real-time applications in the future.

Another trend in CS-based IoMT is the use of machine learning algorithms to improve the reconstruction quality of the signal. The use of deep learning and machine learning algorithms can help to improve the accuracy of the reconstruction while reducing the computational overhead. There should be trade-off between different algorithms with respect to compression ratio and sensing matrices.

Another trend is the integration of CS with other technologies such as blockchain and edge computing. The use of blockchain can enhance data security and privacy, while edge computing can reduce latency and improve real-time decision-making.

7 Conclusion

IoMT applications require a vast array of smart devices with varying capabilities to sense, communicate, and collaborate in order to collect and provide health services. However, a large number of sensors generate redundant data, causing unnecessary network traffic and degrading system performance. Compressed sensing (CS) can reduce the amount of data collected and alleviate pressure on wireless communication. This chapter provides an overview of CS theory, including the physiological signal acquisition model, sensing matrices, and reconstruction algorithms. It also covers the integration of CS techniques in the IoMT platform, divided into three layers. The acquisition layer involves a large number of sensors communicating with a central fusion node, where CS can reduce power consumption. The processing unit can implement edge computing on FPGAs for real-time performance, and several CS reconstruction algorithms have been proposed. The application layer can extend the use of CS to various healthcare applications. The chapter also discusses open issues and future trends in the field of IoMT. The emerging topic is discussed a CS encoder that directly provides a compressed signal acquisition, enabling the next era of digital sampling. The use of structured and deterministic sensing matrices can simplify acquisition operations and allow real-time data reconstruction and feature extraction from compressed measurements. Additionally, CS can be used for secure and encrypted transmission by leveraging the randomness inherent in the acquisition process without requiring additional processing.

Acknowledgments This work was partially supported by project SERICS (PE00000014) under the MUR National Recovery and Resilience Plan funded by the European Union – NextGenerationEU.

References

1. M.M. Islam, A. Rahaman, M. R. Islam, Development of smart healthcare monitoring system in IoT environment. SN Comput. Sci. **1**, 1–11 (2020)
2. P. Sundaravadivel, E. Kougianos, S.P. Mohanty, M.K. Ganapathiraju, Everything you wanted to know about smart health care: evaluating the different technologies and components of the internet of things for better health. IEEE Consum. Electron. Mag. **7**(1), 18–28 (2017)
3. K. Anjali, R. Anand, S.D. Prabhu, R. Geethu, IoT based smart healthcare system to detect and alert covid symptom, in *2021 6th International Conference on Communication and Electronics Systems (ICCES)* (IEEE, Piscataway, 2021), pp. 685–692
4. S. Ahmad, H.A. Abdeljaber, J. Nazeer, M.Y. Uddin, V. Lingamuthu, A. Kaur, Issues of clinical identity verification for healthcare applications over mobile terminal platform. Wireless Commun. Mob. Comput. **2022**, 1–10 (2022)
5. J. Abbasi, L. Alina, A.M. Abro, B. Lal, Sehat: smart e-health app for telediagnosis and first opinion, in *2021 6th International Multi-Topic ICT Conference (IMTIC)* (IEEE, Piscataway, 2021), pp. 1–6
6. H.A. Khan, R. Abdulla, S.K. Selvaperumal, A. Bathich, IoT based on secure personal healthcare using RFID technology and steganography. Int. J. Elect. Comput. Eng. (2088–8708) **11**(4), 3300–3309 (2021)

7. E.J. Candes, T. Tao, Near-optimal signal recovery from random projections: universal encoding strategies? IEEE Trans. Inf. Theory **52**(12), 5406–5425 (2006)
8. C.G. Graff, E.Y. Sidky, Compressive sensing in medical imaging. Appl. Opt. **54**(8), C23–C44 (2015)
9. Y. Oiknine, I. August, V. Farber, D. Gedalin, A. Stern, Compressive sensing hyperspectral imaging by spectral multiplexing with liquid crystal. J. Imaging **5**(1), 3 (2018)
10. Y. Wiaux, L. Jacques, G. Puy, A.M. Scaife, P. Vandergheynst, Compressed sensing imaging techniques for radio interferometry. Mon. Not. R. Astron. Soc. **395**(3), 1733–1742 (2009)
11. A. Roozgard, N. Barzigar, P.K. Verma, S. Cheng, Genomic data privacy protection using compressed sensing. Trans. Data Priv. **9**(1), 1–13 (2016)
12. X. Liu, M. Zhang, T. Xiong, A.G. Richardson, T.H. Lucas, P.S. Chin, R. Etienne-Cummings, T.D. Tran, J. Van der Spiegel, A fully integrated wireless compressed sensing neural signal acquisition system for chronic recording and brain machine interface. IEEE Trans. Biomed. Circuits Syst. **10**(4), 874–883 (2016)
13. K. Hayashi, M. Nagahara, T. Tanaka, A user's guide to compressed sensing for communications systems. IEICE Trans. Commun. **96**(3), 685–712 (2013)
14. H. You, Z. Ma, W. Li, J. Zhu, A speech enhancement method based on multi-task bayesian compressive sensing. IEICE Trans. Inf. Syst. **100**(3), 556–563 (2017)
15. Y. Arjoune, N. Kaabouch, H. El Ghazi, A. Tamtaoui, A performance comparison of measurement matrices in compressive sensing. Int. J. Commun. Syst. **31**(10), e3576 (2018)
16. D. Ramalho, K. Melo, M. Khosravy, F. Asharif, M.S.S. Danish, C.A. Duque, A review of deterministic sensing matrices. Compressive Sens. Healthcare 89–110 (2020)
17. Y. Yang, H. Liu, J. Hou, A compressed sensing measurement matrix construction method based on TDMA for wireless sensor networks. Entropy **24**(4), 493 (2022)
18. H. Mamaghanian, N. Khaled, D. Atienza, P. Vandergheynst, Compressed sensing for real-time energy-efficient ECG compression on wireless body sensor nodes. IEEE Trans. Biomed. Eng. **58**(9), 2456–2466 (2011)
19. M. Rani, S. Dhok, R. Deshmukh, EEG seizure detection from compressive measurements, in *Advances in VLSI, Communication, and Signal Processing: Select Proceedings of VCAS 2018* (Springer, Berlin, 2020), pp. 963–969
20. K. Abualsaud, M. Mahmuddin, R. Hussein, A. Mohamed, Performance evaluation for compression-accuracy trade-off using compressive sensing for EEG-based epileptic seizure detection in wireless tele-monitoring, in *2013 9th International Wireless Communications and Mobile Computing Conference (IWCMC)* (IEEE, Piscataway, 2013), pp. 231–236
21. Z. Zhang, X. Liu, S. Wei, H. Gan, F. Liu, Y. Li, C. Liu, F. Liu, Electrocardiogram reconstruction based on compressed sensing. IEEE Access **7**, 37 228–37 237 (2019)
22. A. Ravelomanantsoa, A. Rouane, H. Rabah, N. Ferveur, L. Collet, Design and implementation of a compressed sensing encoder: application to emg and ECG wireless biosensors. Circuits Syst. Signal Process. **36**, 2875–2892 (2017)
23. M. Balouchestani, S. Krishnan, Effective low-power wearable wireless surface EMG sensor design based on analog-compressed sensing. Sensors **14**(12), 24 305–24 328 (2014)
24. S. Jain, U. Oswal, K.S. Xu, B. Eriksson, J. Haupt, A compressed sensing based decomposition of electrodermal activity signals. IEEE Trans. Biomed. Eng. **64**(9), 2142–2151 (2016)
25. H. Djelouat, X. Zhai, M. Al Disi, A. Amira, F. Bensaali, System-on-chip solution for patients biometric: a compressive sensing-based approach. IEEE Sens. J. **18**(23), 9629–9639 (2018)
26. J. Sheng, C. Yang, M.C. Herbordt, Hardware-efficient compressed sensing encoder designs for wbsns, in *2015 IEEE High Performance Extreme Computing Conference (HPEC)* (IEEE, Piscataway, 2015), pp. 1–7
27. M. Rani, S.B. Dhok, R.B. Deshmukh, A systematic review of compressive sensing: concepts, implementations and applications. IEEE Access **6**, 4875–4894 (2018)
28. R. Manchanda, K. Sharma, A review of reconstruction algorithms in compressive sensing, in *2020 International Conference on Advances in Computing, Communication & Materials (ICACCM)* (2020), pp. 322–325

29. L. Hu, J. Zhou, Z. Shi, Q. Fu, A fast and accurate reconstruction algorithm for compressed sensing of complex sinusoids. IEEE Trans. Signal Proces. **61**(22), 5744–5754 (2013)
30. J.K. Pant, W.-S. Lu, A. Antoniou, New improved algorithms for compressive sensing based on ℓ_p norm. IEEE Trans. Circuits Syst. II Exp. Briefs **61**(3), 198–202 (2014)
31. S. Chen, D. Donoho, Basis pursuit, in *Proceedings of 1994 28th Asilomar Conference on Signals, Systems and Computers*, vol. 1 (1994), pp. 41–44
32. S. Burer, A.N. Letchford, Non-convex mixed-integer nonlinear programming: a survey. Surv. Oper. Res. Manag. Sci. **17**(2), 97–106 (2012) [Online]. Available: https://www.sciencedirect.com/science/article/pii/S1876735412000037
33. A. Vince, A framework for the greedy algorithm. Discrete Appl. Math. **121**(1), 247–260 (2002) [Online]. Available: https://www.sciencedirect.com/science/article/pii/S0166218X01003626
34. A. Movahed, A. Panahi, G. Durisi, A robust rfpi-based 1-bit compressive sensing reconstruction algorithm, in *2012 IEEE Information Theory Workshop* (2012), pp. 567–571
35. A. Draganic, I. Orovic, S. Stankovic, On some common compressive sensing recovery algorithms and applications - review paper. CoRR, abs/1705.05216 (2017) [Online]. Available: http://arxiv.org/abs/1705.05216
36. E. Crespo Marques, N. Maciel, L. Naviner, H. Cai, J. Yang, A review of sparse recovery algorithms. IEEE Access **7**, 1300–1322 (2019)
37. A.L. Machidon, V. Pejović, Deep learning for compressive sensing: a ubiquitous systems perspective. Artif. Intell. Rev. **56**(4), 3619–3658 (2023) [Online]. Available: https://doi.org/10.1007/s10462-022-10259-5 https://link.springer.com/10.1007/s10462-022-10259-5
38. J.Y. Park, M.B. Wakin, A.C. Gilbert, Modal analysis with compressive measurements. IEEE Trans. Signal Process. **62**(7), 1655–1670 (2014)
39. M. Rani, S.B. Dhok, R.B. Deshmukh, EEG seizure detection from compressive measurements, in *Advances in VLSI, Communication, and Signal Processing*, ed. by D. Dutta, H. Kar, C. Kumar, V. Bhadauria (Springer Singapore, Singapore, 2020), pp. 963–969
40. S. Qiu, Z. Li, W. He, L. Zhang, C. Yang, C.-Y. Su, Brain–machine interface and visual compressive sensing-based teleoperation control of an exoskeleton robot. IEEE Trans. Fuzzy Syst. **25**(1), 58–69 (2017)
41. C.-Y. Chou, Y.-W. Pua, T.-W. Sun, A.-Y.A. Wu, Compressed-domain ECG-based biometric user identification using compressive analysis. Sensors **20**(11) (2020) [Online]. Available: https://www.mdpi.com/1424-8220/20/11/3279
42. M. Al Disi, H. Djelouat, C. Kotroni, E. Politis, A. Amira, F. Bensaali, G. Dimitrakopoulos, G. Alinier, ECG signal reconstruction on the IoT-gateway and efficacy of compressive sensing under real-time constraints. IEEE Access **6**, 69 130–69 140 (2018)
43. M. Zhang, H. Zhang, C. Zhang, D. Yuan, Communication-efficient quantized deep compressed sensing for edge-cloud collaborative industrial IoT networks. IEEE Trans. Ind. Inform. **19**(5), 6613–6623 (2022)

Occupancy Prediction in Buildings: State of the Art and Future Directions

Irfanullah Khan, Emilio Greco, Antonio Guerrieri, and Giandomenico Spezzano

1 Introduction

Recently, the need to enhance energy efficiency in buildings while keeping good comfort, security, and safety levels is driving new research areas [21, 51]. Among these, a very promising one regards occupancy prediction. Occupancy prediction refers to the process of forecasting the individuals present in a specific space at any given time. It serves as a pivotal component of intelligent building systems, leveraging diverse technologies such as the Artificial Intelligence (AI) [61], and Internet of Things (IoT) [7] to establish highly efficient and sustainable buildings. Predicting occupancy in building environments heavily relies on IoT technologies, encompassing sensors, cameras, and smart devices, which monitor and gather data about the building's surroundings. Once data is gathered, AI techniques, e.g., Machine Learning (ML) and Deep Learning (DL), come into play to process and analyze this data. AI algorithms can extract valuable insights from historical occupancy data and sensor readings, identifying patterns, correlations, and trends. Through training predictive models, these algorithms enable the forecast of future room occupancy based on the amassed data.

I. Khan
University of Calabria, Rende (CS), Italy

Institute for High Performance Computing and Networking of the National Research Council of Italy (ICAR-CNR), Rende (CS), Italy
e-mail: irfanullah.khan@dimes.unical.it

E. Greco · A. Guerrieri (✉) · G. Spezzano
Institute for High Performance Computing and Networking of the National Research Council of Italy (ICAR-CNR), Rende (CS), Italy
e-mail: emilio.greco@icar.cnr.it; antonio.guerrieri@icar.cnr.it; giandomenico.spezzano@icar.cnr.it

C. Savaglio et al. (eds.), *Device-Edge-Cloud Continuum*, Internet of Things,
https://doi.org/10.1007/978-3-031-42194-5_12

Occupancy prediction in building environments holds multifaceted applications, encompassing energy efficiency, comfort level optimization, green building initiatives, air control management, and security enhancement. One of the primary benefits of occupancy prediction is improving energy efficiency [27]. With occupancy prediction, building systems such as heating, ventilation, and air conditioning (HVAC), lighting, and power distribution can be dynamically adjusted based on anticipated occupancy levels. For instance, if a particular area is projected to have low occupancy, HVAC systems can be scaled back or even temporarily deactivated, reducing unnecessary energy consumption. Similarly, lighting systems can be intelligently controlled to illuminate only occupied areas, reducing electricity usage without compromising safety or comfort. Moreover, occupancy prediction allows for the identification of usage patterns and trends over time. By analyzing historical occupancy data, building managers can gain insights into peak usage hours, seasonal variations, and occupancy trends on different days of the week. Armed with this knowledge, energy management strategies can be fine-tuned to align with specific occupancy patterns, further maximizing energy efficiency and cost savings. Another significant benefit of occupancy prediction is improving occupant comfort [31]. With occupancy prediction, building systems such as temperature control, lighting, and ventilation can be adjusted in real time based on anticipated occupancy levels. This ensures that the indoor environment remains comfortable and conducive to the activities taking place within the building. For example, if a conference room is expected to be occupied, the temperature and ventilation settings can be adjusted accordingly to create a comfortable meeting environment. By actively managing the indoor environment based on occupancy prediction, potential discomfort due to overcrowding or inadequate environmental conditions can be minimized. Spaces can be prepared in advance, ensuring appropriate temperature levels, optimal lighting conditions, and adequate airflow for the anticipated number of occupants. This proactive approach enhances occupant satisfaction, productivity, and overall well-being.

Numerous methodologies have been deployed so far to collect data on the building environment, enabling accurate occupancy prediction. These methodologies encompass a broad spectrum of IoT technologies, including sensors, radars, cameras, and wearable devices [43]. Sensors are perhaps the most commonly used method, with many studies relying on sensors such as carbon dioxide (CO_2), passive infrared (PIR), temperature, humidity, and light to capture data on occupancy patterns. For example, Khan et al. [30] used a combination of different sensors, namely CO_2, light, and humidity, to capture data on occupancy in a room. In addition to sensors, cameras have also been used to capture visual data on occupancy. For instance, Dino et al. [17] used cameras to monitor occupancy in a university building, while Aliero et al. [4] utilized a combination of cameras and sensors to track occupancy in a living room. More recently, wearable devices and radar technology have emerged as potential methods for occupancy prediction. Taniguchi et al. [56] used wearable devices, i.e., smartwatches, to detect door operations which can be used for occupants detection in a home, while Hasan et al. [23] employed ultra-wideband radars technology for counting people walking through a doorway.

These studies demonstrate the range of techniques available for occupancy detection and prediction and highlight the importance of choosing the most appropriate method for the specific context. Each of these techniques has its strengths and weaknesses, and building managers must carefully consider which methods to use based on their specific needs and requirements. By leveraging the data collected by these methods in building environments, Artificial Intelligence techniques can be employed to predict occupancy levels and optimize building operations accurately. Such techniques can also be smartly applied at the edge [5, 10, 50] and, by enabling collaboration with the cloud for augmenting computational power and the IoT devices for collecting data and making a first data elaboration, can reach the so-called Device–Edge–Cloud Continuum [29].

The previous research on occupancy prediction in building environments has utilized various data-driven algorithms, including traditional machine learning and deep learning techniques [54]. The study by Esrafilian-Najafabadi et al. [19] investigated how the choice of occupancy prediction models affects the efficiency of HVAC control systems in residential buildings. They employed four machine learning methods, namely decision trees, k-nearest neighbors, multi-layer perceptron, and gated recurrent units, to predict future occupancy from various perspectives. While ML algorithms have been useful in predicting occupancy levels in buildings, DL techniques provide additional benefits by facilitating the processing of complex datasets with high accuracy. Mohammadabadi et al. [39], for example, proposed a novel approach to predict occupancy using indoor climate data through a DL method named CNN-XGBoost. The authors also compare the performance of this proposed method with other supervised and unsupervised ML algorithms, as well as artificial neural network algorithms. All the studies above demonstrate the effectiveness of ML and DL for occupancy prediction and highlight the potential for these methods to enhance building performance, comfort, and energy efficiency.

The recently published research has concentrated on different methods and techniques for the prediction of room occupancy in building environments and has achieved promising outcomes. However, further investigation is required to enhance and fine-tune these methods and techniques. This chapter aims to present the current state of the art by providing an overview of the main recent publications published in the field of occupancy prediction in building environments. In particular, we will begin by exploring the main IoT technologies used to monitor presence in buildings. Such technologies play a crucial role in assessing occupancy and enabling efficient management of building spaces. Once we have established an understanding of the monitoring methods, we will move on to examine machine learning and deep learning algorithms employed for predicting occupancy. These advanced computational techniques have shown great promise in accurately forecasting the number of individuals in building environments. By leveraging historical data and patterns, these algorithms can learn and make informed predictions, contributing to effective space utilization and resource allocation. Finally, we will explore some future directions within this field, mainly regarding the focus on the prediction of the exact number of people in buildings, the match of occupancy prediction with localization, the activity forecasting, and the use of approaches based on the edge computing, in general, and Device–Edge–Cloud Continuum, in particular, for occupancy prediction, such as the Federated Learning (FL).

The organization of this chapter is as follows: In Sect. 2, various techniques used for monitoring building environments are discussed. Section 3 focuses on the AI techniques used to analyze collected data and forecast room occupancy. Section 4 examines potential future directions for occupancy prediction. Finally, Sect. 5 concludes the chapter.

2 Monitoring Methods

The utilization of data-driven techniques to predict room occupancy has gained significant popularity due to its ability to optimize building management and enhance energy efficiency. To achieve this, it is crucial to monitor the environment using a variety of devices classified into four distinct categories: sensors, cameras, wearables, and radars. Sensors are employed to detect changes in the environment, while cameras analyze video feeds to identify human shapes or movements. Radars emit radio waves that bounce off various objects, including humans, allowing for the detection of reflections. Finally, wearable technologies utilize advanced technology to ascertain the presence of individuals within a room. By amalgamating data collected from these diverse monitoring devices, a more comprehensive and precise model can be developed to predict room occupancy accurately. This, in turn, facilitates optimization of building management, energy efficiency, cost reduction, as well as improvements in safety and security. In the following subsections, we will delve into the details of each monitoring method and carefully examine its respective strengths and limitations, which will also be described in Table 1. Moreover, Fig. 1 will show some example devices among the ones used for occupancy detection/forecasting.

2.1 Sensors

The use of sensors for monitoring the room environment is a critical aspect of predicting room occupancy since it is a widely adopted and uncomplicated method that can be easily installed in a building. By collecting data on factors such as CO_2 concentration, temperature, light, humidity, and noise, sensors can provide valuable insights into whether a room has any occupancy or not. Also, radio signals can be treated as advanced sensors by considering their received strength in specific places. Each of these sensors has a unique approach to monitoring the environment of a room. For example, temperature sensors can detect the heat generated by people in a room, while CO_2 sensors can detect the concentration of carbon dioxide produced by their breath. By analyzing these sensor data points in real time, a machine learning model can learn the patterns and correlations between different

Table 1 Overall literature of monitoring method for occupancy prediction in buildings

Method	Sensor type	Ref.	Advantage	Limitation	Performance evaluation	Application	Building type
Active sensor	Ultrasonic sensor	[22]	Non-intrusive in nature,	Comparatively expensive,	Accuracy is above 90%	Occupancy counting, activity recognition	Room
	Ultrasonic sensor	[38]	real-time monitoring	needs power source to operate,	Not declared	Occupancy counting, occupancy detection	Classroom
	Ultrasonic sensor	[13]	high accuracy		Improved accuracy from 85% to 93%	Calculate the position and the orientation of a robot	Room
	Bluetooth beacons	[46]	Non-intrusive in nature,	Reliance on user devices,	Ranging from 92% to 99% F-measure	Office multi-occupancy detection	Office
	Bluetooth beacons	[15]	real-time monitoring	limited range,	Accuracy = 97.97%	Low cost room occupancy prediction	Room
	Bluetooth beacons	[9]	low cost	security concerns	Accuracy range from 88% to 92%	Human positioning in a room	Room
	Wi-Fi technology	[41]	Non-intrusive in nature, and real-time monitoring	High cost, privacy concern	84.6% accuracy	Classroom occupancy	Classroom
	Wi-Fi technology	[40]			96% accuracy	Counting occupancy number	Room
	Active RFID	[37]	Non-intrusive and real time	High cost, security concerns	100% presence in office accuracy	Thermal zone occupancy	Office

(continued)

Table 1 (continued)

Passive sensors	PIR sensor	[52]	Easy installation,	False positive result,	Accuracy >99%	Occupancy detection	IT hub
	PIR sensor	[49]	low maintenance,	limited range	30% load reduction	Building energy management system	Not declared
	PIR sensor	[36]	non-intrusive in nature		Not declared	Residential houses	Room, and buildings
	CO_2	[55]	Non-intrusive,	Slow response time,	Reducing energy up to 51.4%	Controlling the HVAC system	Campus classrooms
	CO_2	[58]	useful for calculation	limited accuracy,	Accuracy = 95.48%	Tracking occupancy in a room	Room
	CO_2	[42]	number of occupancy	and expensive	RMSE range 1.8–1.9	Office building's occupancy prediction	Office
	Sensor combination	[20]	Adaptability, improve accuracy	Increased complexity, relatively expensive	Accuracy from 87–99.5%	Occupancy detection	Building
	Sensor combination	[30]			Accuracy is 94.5%	Occupancy detection	Room
Camera	Camera	[57]	High accuracy, real time monitoring, provide comprehensive data, easy implementation	Expensive, privacy concern,	Accuracy = 80.62%	Reduce energy consumption	Building
	Camera	[60]			Not declared	Positioning system	Building
	Camera	[25]			MAE = 0.94, MSE = 1.35	Number of occupants	Building
	Camera	[11]			RMSE = 0.883, NRMSE = 0.141	Occupancy counting	Offices
	Camera	[17]			MAE = 0.04	Estimate number of people	Large, crowded space
UWB Radar	UWB Radar	[26]	Privacy friendly, non-intrusive in nature, real-time monitoring, low power consumption	Expensive, complex signal processing, limited range	Not declared	Occupancy prediction	Buildings
	UWB Radar	[23]			Accuracy = 90%	Occupants counting	Door
	UWB Radar	[8]			Accuracy = 83%	Number of occupants	Room
Wearable technology	Wearable technology	[2]	Real-time data collection, non-intrusive in nature	Expensive, limited range	Not declared	Occupants comfort	Building
	Wearable technology	[44]			Not declared	Occupants comfort	Building

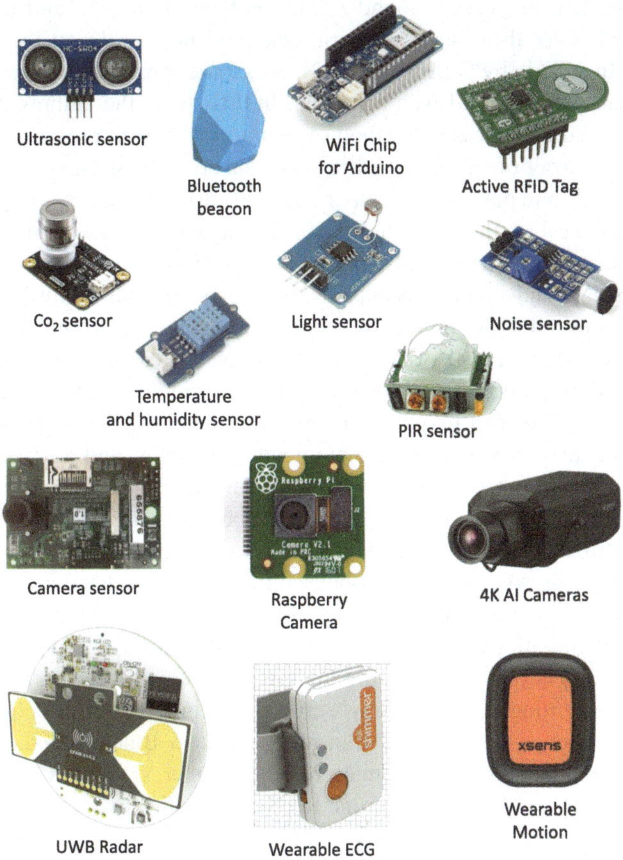

Fig. 1 Some example devices among the ones used for occupancy detection/prediction

environmental factors and room occupancy. This study categorizes the sensors used for monitoring room occupancy into two main categories: active sensors and passive sensors. In the following, we will provide a brief overview of both the categories.

2.1.1 Active Sensors

Active sensors have gained considerable attention in recent years for their ability to detect room occupancy accurately. Active sensors emit a signal, which interacts with the environment to provide information about the presence or movement of people in the room. Active sensors require a power source to function and can be used to provide real-time information about occupancy. Active sensors are further categorized into subcategories, namely ultrasonic sensors, Bluetooth beacons, and Wi-Fi, each with unique advantages.

Ultrasonic sensors release sound waves of high frequency and calculate the duration it takes for the waves to return, enabling them to detect the presence of individuals. In a study by Ghosh et al. [22], ultrasonic sensors were used to monitor room occupancy in a data-driven approach. In this study, the authors introduced a non-intrusive and non-invasive method of monitoring human presence and activity by utilizing an array of dissimilar ultrasonic sensors. The sensors are arranged in two distinct grids and placed on the door frame in various locations. The system can detect a series of events that occur as individuals engage in various primary, postural, and group activities. Additionally, it can recognize different indoor movements, room occupancy, and identify occupants. Similarly, another study [38] also uses this type of sensor for the prediction of occupants in a classroom. In this article, the authors describe an IoT system intended to assist teachers in complying with COVID-19 safety guidelines during classroom instruction. The system is capable of counting the number of people in the classroom and their spatial distribution. It is made up of three components: a Gate node, IoT nodes, and a server. The Gate node, which is positioned at the entrance, can monitor the number of individuals entering and leaving the room by detecting door crossings. Ultrasonic distance sensors are utilized as IoT nodes and are positioned on seats to track seat occupancy. While ultrasonic sensors are commonly used for occupancy detection, they do have some limitations, including line-of-sight requirements, limited range, sensitivity to environmental conditions, difficulty in object differentiation, and blind spots.

Recent studies have been exploring the use of Bluetooth beacons for occupancy detection, highlighting the technology's crucial role in this area. Bluetooth beacons are small, low-power devices that transmit signals to Bluetooth-enabled devices within a certain range, such as smartphones or other sensors. Using Bluetooth beacons makes it possible to determine the location of individuals and track their movements within a space. This information can then be used to predict occupancy levels in a particular area. For example, Pratama et al. [46] examine how mobile phone power consumption and beacon signals can indicate human presence in multi-user office spaces. The hypothesis is that combining these sources of information can enhance the accuracy and level of detail in detecting occupancy. The study applies various techniques (i.e., machine learning, optimization, and probabilistic) to test this hypothesis and evaluates the effectiveness of each source and its fusion in a real office setting. Finally, the study compares the performance of each individual method of detecting occupancy. Similarly, another study also proposes a method for detecting the presence of humans in multi-user environments based on variations in Bluetooth Low Energy (BLE) signals [15]. The method is tested on five different datasets using pattern recognition models and compared to IEEE 802.11 (Wi-Fi) systems. Results show an average accuracy of 97.97% for occupancy classification and an average difference of 0.32 persons for estimating the number of people in a room. Another study also explores using Bluetooth beacons to collect statistics about room occupancy in healthcare monitoring [9]. A new spatial feature is introduced that can be applied to any configuration, allowing differentiation between positioning information and the presence of residents. The study shows promising results from experiments with various floor plans. Bluetooth technology is a cheap data collection source, but it has some limitations. For example, its range is limited, typically up to 30 feet (10 meters), and it can only connect a few devices at a time. On the other hand, Wi-Fi offers higher speed and greater range than Bluetooth, making it ideal for detecting and predicting room occupancy.

Wi-Fi technology has gained popularity in the last decade for detecting room occupancy because it is easy to install and provides real-time data on the presence of Wi-Fi connected devices in a room. Wi-Fi sensing infrastructure comprising multiple Wi-Fi access points (APs) is installed in the room. The APs collect metadata about the Wi-Fi-connected devices in the room, such as their signal strength and time spent connecting to each AP. This metadata can be analyzed using machine learning algorithms to predict real-time occupancy levels. The study by Mohottige et al. [41] focused on utilizing machine learning models to estimate classroom occupancy using Wi-Fi sensing infrastructure. Their research analyzed metadata from a dynamic wireless network with numerous access points to identify obstacles in predicting room occupancy. The authors proposed a technique to map access points to classrooms automatically and evaluated various algorithms to enhance the accuracy of occupancy estimation. The study models classroom occupancy and evaluates algorithms such as Logistic Regression and SVM. Similarly, Mohammadmoradi et al. [40] present a method for tracking room occupancy using wireless and light sensing technologies in doorways to track entrance of the room. The proposed solution is inexpensive and unobtrusive and utilizes ultra-wideband and narrowband wireless signals for sensing. The possibility of using existing Wi-Fi infrastructures for people counting is also evaluated. The method has been tested and evaluated in different scenarios, showing promising results.

Passive and Active RFID tags are also used in the literature for occupancy detection systems. RFID is a technology that uses radio waves to transmit data between RFID tags and one or more readers. While passive RFID tags have a short communication range since they have no batteries, but they rely on the energy transmitted by the reader to power the tag and facilitate communication, the active RFID tags have their own power source (usually a battery) and actively transmit data signals to the reader. For this reason, active RFID tags are the most commonly used for occupancy detection in buildings. As an example, the authors in [37] proposed an active RFID-based occupancy detection system that enables demand-driven HVAC operations. The system aims to detect and track multiple stationary and mobile occupants in multiple spaces simultaneously. By estimating the thermal zone of each occupant, the proposed system provides real-time reports on the number of occupants in each thermal zone.

Active sensors are useful for detecting movement but may not always provide an accurate picture of room occupancy. In particular, where there are high levels of radio wave interference, these technologies can experience issues and, as a result, perform poorly. In these cases, passive sensors can be used as an alternative approach to detect and, consequently, predict room occupancy.

2.1.2 Passive Sensors

Passive sensors are sensors that detect the presence of objects or individuals without actively emitting any signals. They rely on detecting changes in the environment, such as changes in temperature, light, or motion, to determine the presence of an object or individual. Examples of passive sensors include infrared sensors, motion detectors, and light sensors.

The most common example of a passive sensor used for occupancy prediction is the PIR sensor. PIR sensors detect the infrared radiation emitted by the human body and can be used to detect the presence of people within the room. A study by Stamatescu et al. [52] used a PIR sensor and thermopile sensor and evaluated the performance of random forests, an ensemble model, in predicting individual cell temperature values and detecting occupancy status. The method's effectiveness was assessed through a real-world case study deployed in an IT Hub in Bucharest, where over three weeks of ground data was analyzed and utilized to predict room occupancy. Similarly, another paper has also used a PIR sensor to detect the presence of occupants to enhance the energy efficiency of a building [49]. This paper investigates real-world data streams from numerous digital PIR occupancy sensors that are integrated into LED lighting fixtures in a recent IoT deployment in a California building. Initially, a data-driven approach is developed to detect anomalies in these data streams. The findings are subsequently utilized to improve energy efficiency in the building and create prospects for providing demand response services. Additionally, another study involves collecting data from PIR sensors installed in four residential houses in the USA and proposing an inhomogeneous Markov model for predictive building controls [36]. To predict occupancy presence accurately, the study employs change-point analysis of historical data to determine the optimal training periods. The research assesses several scenarios utilizing 15-minute data to predict occupancy presence at different time intervals, evaluating both room-level and house-level predictions. Although PIR sensors are a proven method for predicting room occupancy, the study acknowledges that other techniques are available and may require further investigation.

CO_2 is another method that is becoming increasingly popular as a means of gathering data to estimate room occupancy. In this regard, the study developed a systematic approach to regulate the HVAC system in a campus classroom, which relied on the concentration of CO_2 in the environment [55]. To predict CO_2 levels, the researchers selected and fine-tuned six advanced machine-learning algorithms and found that the multilayered perceptron network was the most effective due to its ability to learn nonlinear patterns in the CO_2 data. Compared to traditional ON/OFF controllers, model-based control systems demonstrated better control accuracy and energy efficiency, as evidenced by the study's findings. Similarly, in this article [58], the authors propose a novel method of monitoring room occupancy by merging decentralized building automation technology with a software application gateway that uses visible light communication. The article also introduces a new software application gateway that uses the MQTT protocol to facilitate interoperability between its technology and the IBM Watson IoT platform. By creating an interoperable hardware gateway that takes advantage of artificial neural network mathematical techniques for CO_2 waveform prediction, the researchers leveraged building automation technology. In addition, Motuziene et al. [42] make use of Extreme Learning Machine (ELM) models with two different optimization algorithms, Genetic (GA-ELM) and Simulated Annealing (SA-ELM), to forecast occupancy in an office building based on measured CO_2 levels. Both models demonstrate comparable and high levels of accuracy in prediction, with values ranging from 0.73 to 0.74 and RMSE values ranging from 1.8 to 1.9 over the entire measurement period. While CO_2 data is beneficial for estimating room occupancy, it may exhibit some delays in accurately reflecting changes in occupancy levels. In addition to CO_2 data, it can be advantageous to investigate other passive sensors for estimating room occupancy. It is beneficial to integrate multiple sensors to address the limitations of each sensor and achieve better outcomes.

A combination of different sensors is used to explore the goodness of the passive sensors, including light sensor, humidity sensor, temperature, and CO_2 sensor for the prediction of room occupancy. In this regard, the study determined the occupancy of a building by utilizing easily measurable parameters of the indoor environment, such as light, temperature, relative humidity, CO_2 level, and humidity ratio [20]. A data-driven approach to detecting occupancy was developed using an artificial neural network (ANN) based on particle swarm optimization (PSO) and implemented. The proposed approach was validated with 17 different models using a measured dataset. A satisfactory result was achieved with the combination of these sensors. Another study [30] used different sensor data such as light, humidity, humidity ratio, temperature, and CO_2 concentration to detect the presence and absence of the occupants. The study used LSTM neural network to train the model on these data and combine it with federated learning to ensure the privacy of the data being trained at edge nodes. The model achieved very high accuracy to show the goodness of the proposed approach.

Passive sensors have several advantages, including their low cost, low power consumption, and easy installation. However, they are susceptible to false alarms caused by factors such as pets or temperature changes. Passive sensors can be used in various applications, such as building automation systems, energy management, and security systems. Overall, passive sensors offer a reliable and cost-effective solution for occupancy detection in various indoor environments. A detailed overview of the possible advantages and limitations of both active and passive sensors is described in Table 1.

2.2 Cameras

Camera-based occupancy detection represents a very accurate method compared to traditional methods such as sensors or manual counting. By detecting occupancy with very good accuracy, also the prediction of the presence can reach good levels. In the literature, all the works related to occupancy detection based on cameras capture real-time images and video data that are analyzed by advanced algorithms to detect and predict the number of people in a room. For example, a study published recently presents a novel approach using deep learning and vision-based technology to detect people and identify and classify the activities of individuals within building spaces [57]. The collected data can be integrated with building energy management systems to create occupancy heat emission profiles, which can then be used to reduce energy loads from heating, ventilation, and air conditioning (HVAC) systems and improve indoor environmental conditions. This approach aims to provide an efficient and effective means of managing building energy consumption while maintaining optimal occupant comfort. Similarly, another research study presents a novel system for indoor positioning that uses an RGB-D camera [60]. This system utilizes multiple cameras to capture depth and color images from various viewpoints and incorporates a data fusion algorithm based on human pose estimation. The authors' study conducted steady and dynamic accuracy evaluations in a multifunctional room to demonstrate the practicality and effectiveness of the proposed system. In another research paper, the authors propose a deep learning-based technique for enhancing building occupancy detection using CCTV cameras [25]. This technique is composed of two main components: feature extraction and three-stage occupancy detection. The feature extraction component uses a deep convolutional neural network to extract shallow and semantic features and generate feature pyramids using a bidirectional feature network. The three-stage occupancy detection component utilizes three detectors with increasing Intersection over Union (IoU) thresholds to identify occupants. The study conducts empirical experiments using CCTV videos from a university building, which shows the technique's superior detection performance compared to baseline models. Finally, another study [17] utilizes a camera to estimate room occupancy using a deep vision-based technique. This research has two objectives: first, to assess the effectiveness of deep vision-based occupancy counting, and second, to implement a deep vision-based occupant-centric control (OCC) strategy in real-world settings [11]. The study evaluates the performance of the occupancy counting method in six offices and implements a deep vision-based energy recovery ventilator control strategy in a small office. The performance of this strategy is then compared with traditional control methods, considering indoor air quality and energy consumption.

Camera-based occupancy prediction is a promising technology that can improve building management and energy efficiency by accurately predicting the number of occupants in a building or a particular area. This technology is particularly useful for large buildings or public spaces where manual counting may be difficult or impractical. However, there is a growing concern over the privacy implications of camera-based occupancy detection. The use of cameras for surveillance and monitoring raises ethical and legal questions, particularly if the images are being used to identify or track individuals. Therefore, it is essential to balance the benefits of this technology with the need to protect individual privacy rights.

2.3 UWB Radars

Ultra-Wideband (UWB) radar [26] is a promising technology that provides an accurate and non-intrusive method of detecting/predicting occupancy levels. Since UWB radars are highly accurate and can detect small movements, they are ideal for room occupancy estimation and forecasting. UWB radar systems typically operate at frequencies between 3.1 and 10.6 GHz and can be used in a variety of applications, including automotive sensing, security, and medical imaging.

A study recently published has utilized UWB radar to count the occupants in a room. In this article [23], the authors introduce a people counting system that utilizes Impulse Radio Ultra-Wideband (IR-UWB) radars to count individuals passing through a doorway. The system incorporates two IR-UWB radars placed horizontally apart to establish a delay effect when a person walks between them, which enables the system to detect movement direction and determine the number of people in a room. The system is capable of real-time people counting and can also be used on saved data, providing flexibility for practical applications. Similarly, another paper proposes a solution for identifying room occupancy using UWB radar, based on principal component analysis (PCA) [8]. The system was tested with up to two subjects within the radar's detection range and achieved 100% accuracy in determining room occupancy. Additionally, the PCA-based approach tried to identify the number of people in the room. When tested for predicting the number of occupants (i.e., one or two occupants) in the room, the algorithm was able to determine the occupancy with 81% and 83% accuracy, respectively.

With the increasing demand for smart buildings and energy-efficient systems, the use of UWB technology for room occupancy detection can be a potential solution. However, as UWB technology is relatively new in this field, more research and experimentation are needed to explore its capabilities and limitations fully. By further exploring and developing UWB-based occupancy detection systems, building owners and operators can potentially benefit from accurate and real-time occupancy information, allowing them to optimize energy consumption and improve overall building management. Therefore, continued exploration and development of UWB technology in this context can lead to significant advancements in smart building technologies.

2.4 Wearable Technology

Wearable technology pertains to electronic devices that can be worn on the human body, either as accessories or integrated into clothing. These devices are typically equipped with sensors, microcontrollers or microprocessors, and wireless connectivity, allowing them to collect data and communicate with other devices. Wearable technology has become increasingly popular in recent years, with applications ranging from fitness tracking to healthcare monitoring.

One potential use of wearable technology is for room occupancy estimation and prediction. Wearable devices can be equipped with sensors that detect the wearer's location and movement patterns, allowing them to track the wearer's presence in a room. This data can be aggregated across multiple wearers to provide an accurate estimation of room occupancy levels. Wearable technology-based occupancy prediction has the advantage of being non-intrusive, as it does not require cameras or other sensors to be placed in the room itself. For example, the paper in [2] investigates the utilization of wearable devices for occupancy detection and for measuring and monitoring the thermal comfort of building occupants. The authors developed a mobile application that gathers data from sensors in cell phones and wearable devices, such as air temperature, relative humidity, skin temperature, heart rate, perspiration rate, and location. They conducted a pilot study that used artificial neural networks to estimate the PMV index accuracy of the gathered data in specific locations. The findings reveal the potential for developing new models to predict the PMV index using the parameters captured by wearable devices. Similarly, another paper studied thermal comfort conditions in two public buildings in Croatia using wearable sensors and standard equipment [44]. Eight occupants participated, located in two different climate zones. Artificial neural networks were used to model metabolic responses, with a validation overlap of 90%.

Wearable technology, such as smartwatches or fitness trackers, can be useful tools for estimating room occupancy, as they can track the movement and activity of individuals within a space [1]. However, this technology is relatively new, and there may still be some limitations or challenges that need to be addressed. For example, the accuracy of the data collected by wearable devices may vary depending on factors such as the type of sensor used, the placement of the device on the body, and the individual's activity level. Additionally, there may be concerns about privacy and data security when using wearable technology in shared spaces. Further exploration and research are needed to fully understand wearable technology's potential benefits and limitations for estimating room occupancy.

3 Data Analysis for Occupancy Prediction

After gathering data through sensors, cameras, radars, wearables, and so on, analyzing it is an important aspect of building management, in general, and occupancy prediction, in particular. Data analysis can also help to optimize energy usage, maximize occupant comfort, and improve security and safeness in buildings [12], but these aspects are out of scope for this chapter.

Once we have gathered data using appropriate methods, we can then apply data-driven techniques to analyze the data and predict building occupancy. In the realm of data analysis, data-driven techniques play a critical role in uncovering patterns, relationships, and insights. When it comes to occupancy prediction, it is essential to focus on the techniques that have gained significant attention and are widely employed in the existing literature. Among these techniques, machine learning and deep learning stand out as the most important and extensively utilized methods for occupancy prediction. In the following, we will analyze some works on occupancy prediction based on both ML and DL. Moreover, Table 2 will show a fast comparison of the reviewed papers.

3.1 Machine Learning

Machine learning [14] is a specific field within artificial intelligence that concentrates on developing algorithms and statistical models, enabling computer systems to learn from data and improve their performance on a particular task without the need for explicit programming. In the domain of predicting occupancy in smart buildings, machine learning algorithms can be trained on data collected from different sources, which can be sensors, cameras, radars, and/or wearables, to predict the occupancy in an environment. For instance, historical data can be utilized to train machine learning algorithms to recognize patterns of occupancy in a building, such as the number of individuals present in a specific room or area at a particular time of the day. Once trained, these algorithms can predict occupancy levels in real time based on the historical data collected.

Various machine learning techniques have been employed to predict building occupancy, as demonstrated in several papers. For instance, Jin et al. [28] present a novel approach for accurately forecasting building occupancy. More specifically, the proposed approach combines temporal-sequential (TS) analysis and machine learning techniques to forecast occupancy. The researchers leverage hourly occupant data from 16 different buildings to demonstrate the effectiveness of their approach. The TS analysis takes into account the temporal and sequential nature of building occupancy, including seasonal patterns, by incorporating a one-week seasonal period. Additionally, an artificial neural network structure is employed (referred to as TS-week-ANN) to improve predictive performance. Through their experiments, the authors compare the proposed TS-week-ANN approach with other baseline methods. The results reveal that the TS-week-ANN method outperforms the baselines in terms of accuracy and predictive performance. Ryu et al. in [48] develop models driven by indoor environmental data, effectively detecting and predicting occupancy states. The researchers conducted experiments in the Building Integrated Control Test-bed at Dankook University, where they gathered accurate occupancy profiles, as well as data on indoor and outdoor CO_2 concentrations and electricity consumption from lighting systems and appliances. The findings indicate that the proposed models, employing decision tree and hidden Markov model (HMM) algorithms and driven by indoor environmental data exhibit effectiveness in detecting current occupancy states and predicting future occupancy states, respectively. Peng et al. [45] explore energy efficiency by considering occupants' behavior, utilizing machine learning to optimize HVAC settings and achieve significant energy savings. The study focuses on stochastic behavior patterns of occupants in an office building and proposes a demand-driven control strategy that adjusts HVAC

Table 2 The machine learning and deep learning algorithms used in the analyzed literature for occupancy prediction

Method	Algorithm used	Ref	Technique used	Performance Evaluation	Application	Building Type
Machine Learning	TS-week ANN	[28]	Not declared	RMSE = 0.085	Occupancy prediction	Building
	Decision tree and HMM	[48]	CO_2 concentration	Accuracy = 93%	Occupancy detection	Building
	KNN algorithm	[45]	Temperature	Energy saving is between 7–52%	Cooling control	Building
	KNN, SVM, and ANN	[59]	Sensor combination	RMSE = 0.883, NRMSE = 0.141	Occupancy prediction	Offices
	KNN, SVM, RF, and ANN	[62]	Image-based	Accuracy = 95.68%	Occupancy prediction	Building
	Support vector regression	[6]	CO_2 concentration	Accuracy = 61.30%	Occupancy counting	Room
Deep Learning	CNN-XGBoost	[39]	Sensor combination	Accuracy = 90%	Occupancy prediction	Residential buildings
	LSTM	[53]	CO_2 concentration	Not declared	Occupancy prediction	Room
	RNN and LSTM	[32]	Occupancy pattern	MAE = 7.66	Occupancy prediction	Hall
	LSTM and ARIMA	[47]	Wi-Fi network	RMSE = 0.93	Number of occupants	Building
	LSTM	[24]	Sensor combination	RMSE = 0.08397	Occupancy prediction	Office
	LSTM	[30]	Sensor combination	Accuracy = 94.5%	Occupancy detection	Office

settings in real time. The learned occupancy information is then used to optimize temperature setpoints for the cooling system, resulting in significant energy savings. The proposed strategy was tested in various office spaces, demonstrating energy savings ranging from 7% to 52% compared to conventional cooling systems. By reducing the need for human intervention, this approach has the potential to enhance HVAC system efficiency and reduce energy consumption. Similarly, Wang et al. [59] compare machine learning algorithms highlighting the enhanced robustness of occupancy prediction with fused data. They compare three popular machine learning algorithms (k-nearest neighbors, support vector machine, and artificial neural network) and three data sources (environmental data, Wi-Fi data, and fused data) to optimize occupancy prediction models in various scenarios. Evaluation metrics such as mean average error, mean average percentage error, and root mean squared error are used to assess model accuracy. Through an on-site experiment, it was found that the artificial neural network model with fused data performs best, while the support vector machine model is suitable for Wi-Fi data. The study also reveals that the fused dataset enhances robustness in occupancy prediction, even though it may not significantly improve accuracy compared to independent data sources. Incorporating surveillance videos, Yang et al. [62] utilize Convolutional Neural Network-based density estimation methods to extract occupancy information, improving accuracy in building energy simulation. Starting from this information, the study uses four AI algorithms, i.e., K-nearest neighbor, support vector machine, random forest, and artificial neural network for occupancy prediction, and conducts a case study in educational buildings to assess performance. The results indicate high accuracy and low error rates in occupancy prediction. The study successfully reduces the deviation between predicted and actual occupancy schedules, leading to improved accuracy in building energy simulation. Finally, Arief et al. [6] propose a domain adaptation model for Carbon Dioxide–Human Occupancy Counter Double Plus (DA-HOC++) to accurately predict the number of people in a room without adequate labeled data. The proposed semi-supervised domain adaptation method is compared with two baseline methods: Support Vector Regression and Seasonal Decomposition for Human Occupancy Counting. Domain adaptation is useful when historical data is unavailable, and the previous working model can be adapted to the new environment. The proposed method is evaluated using a real-world dataset and is compared with state-of-the-art techniques.

Traditional machine learning methods have shown promising results with high accuracy in predicting room occupancy. However, these techniques have limitations when dealing with complex and large datasets. To overcome these limitations, modern techniques such as deep learning should need to be explored in the context of room occupancy predictions. Deep learning techniques, such as neural networks and convolutional neural networks, have shown superior performance in many fields of study, and by applying these modern techniques to the prediction of room occupancy, more accurate and efficient results can be achieved. Furthermore, deep learning techniques can handle large amounts of data and can identify complex patterns, which are essential in the smart building environment where numerous factors can affect room occupancy, such as temperature, humidity, and lighting.

Therefore, exploring deep learning techniques can open up new opportunities for improving room occupancy prediction in smart building environments.

3.2 Deep Learning

Deep learning [35] is a sub-field of machine learning that employs neural networks to acquire knowledge from data. Differing from conventional machine learning approaches, deep learning algorithms have the ability to automatically extract significant features from raw data, eliminating the necessity for manual feature engineering. This characteristic makes deep learning highly advantageous for intricate tasks such as speech and image recognition, natural language processing, and occupancy prediction [3]. In the context of intelligent building environments, deep learning can be leveraged to enhance the accuracy of room occupancy detection by learning from extensive sensor data. By training a deep neural network using sensor data encompassing temperature, humidity, light, CO_2, and sound levels from diverse areas of a building, it becomes possible to predict occupancy in different rooms at different times of the day. This capability offers potential benefits for optimizing building energy efficiency, as occupancy prediction can inform decisions pertaining to heating, cooling, and lighting in various rooms.

The paper in [39] presents a new deep learning method called CNN-XGBoost for occupancy prediction based on indoor climate data. The study compares the new method with various supervised and unsupervised machine learning and artificial neural network algorithms. Indoor climate data collected from a residential building in Copenhagen over 13 days in December 2021, including CO_2, relative humidity, and temperature, are used in the study. The CNN-XGBoost method comprises a convolutional neural network and a scalable end-to-end tree-boosting classifier. The study evaluates the algorithms using mean absolute error, confusion matrix, and F1 score. The results show that CNN-XGBoost outperforms other algorithms in predicting occupancy levels in all rooms of the test building. In [32], the authors propose a solution for occupancy prediction in a large exhibition hall. The authors propose dividing the hall into sections and using a recurrent neural network (RNN) with long short-term memory (LSTM) units to make accurate occupancy predictions. The effectiveness of the RNN model is evaluated by predicting short-term and long-term occupancy based on historical patterns of occupancy changes in different zones of the hall. The study in [47] aimed to compare two prominent time series prediction methods, LSTM and ARIMA models, for forecasting the number of occupants in smart building environments using Wi-Fi networks. The results showed that the combined LSTM approach outperformed individual models at different time scales, reducing the required number of neurons by 67.13%–74.48%. Furthermore, LSTM forecasts exhibited higher accuracy compared to traditional ARIMA models, with an RMSE reduction of 80.9%–93.4%. Another paper in [24] introduces an IoT framework for capturing indoor environmental parameters to collect occupancy multivariate time series data. The focus is on utilizing the

LSTM network to detect the presence of individuals based on the collected data. An experiment is conducted in an office room to demonstrate the successful implementation of the system for acquiring, processing, and storing environmental information. The obtained data is then compared with other machine learning algorithms such as Support Vector Machine, Naïve Bayes Network, and Multilayer Perceptron Feed-Forward Network. The results indicate that LSTM outperforms the other algorithms in the proposed application based on parametric calibrations. The paper in [18] introduces a rule-based HVAC control system that utilizes deep learning for estimating preconditioning time in residential buildings. The system utilizes occupancy, temperature, and weather data to enhance financial performance, energy efficiency, and occupants' comfort. The proposed approach assumes perfect occupancy prediction and outperforms conventional methods. The optimal control system is selected using the TOPSIS method. The system's sensitivity to temperature setbacks is evaluated, and limitations of occupancy-based control systems are identified. The findings offer valuable insights for improving HVAC control strategies in residential buildings. Stjelja et al. [53] address practical implementation and scalability challenges in predicting occupancy, emphasizing the importance of comprehending performance metrics and introducing a transfer learning method. In particular, they use LSTM networks to forecast occupancy in buildings. Moreover, they introduce a transfer learning method to overcome challenges related to adapting models to different buildings and maintaining robustness in changing conditions. The research shows that ground truth data collection can be significantly reduced to just two days. Finally, in [30], the authors propose a unique method for predicting occupancy in multiple rooms of a smart building that combines FL and LSTM neural networks. The study highlights the importance of FL in predicting occupancy for rooms where training a specific model is not feasible. Moreover, to the best of our knowledge, such a work introduces the first development of a Device–Edge–Cloud Continuum system for occupancy prediction. Through simulations on an existing dataset, the results demonstrate the promising performance of the proposed approach, achieving an accuracy of approximately 0.945.

AI techniques, particularly deep learning, have demonstrated great potential for accurately predicting room occupancy in buildings. By analyzing data on occupancy patterns and other environmental factors, AI models can help building managers optimize heating, cooling, and lighting systems to reduce energy waste and provide a more comfortable environment for occupants. Additionally, AI models can help building managers detect anomalies and identify areas for improvement, allowing them to take proactive measures to optimize building performance. To overcome these limitations, researchers and practitioners are exploring new approaches and techniques that integrate AI with other technologies and developing more sophisticated AI models that can adapt to changing conditions. Addressing privacy concerns and ensuring transparency and accountability in decision-making are also important considerations.

4 Future Direction

In the future, occupancy prediction in buildings is expected to progress in several areas, including the prediction of the exact number of people in buildings, the match of occupancy prediction with localization, activity forecasting of occupants, and the use of approaches based on the edge computing for occupancy prediction, such as the Federated Learning.

Multi-occupancy prediction involves simultaneously predicting the number of occupants in different areas of a building, considering the specific usage patterns of diverse spaces. Occupant localization forecasting focuses on accurately determining the exact positions of individuals within a building, enabling personalized services, optimized resource allocation, and enhanced safety measures. Activity forecasting of occupants aims to predict the specific activities individuals will be engaged in, such as working, resting, or socializing, facilitating tailored environmental adjustments and improved energy efficiency. Moreover, FL can be employed for occupancy prediction, allowing models to be collaboratively trained across multiple buildings without sharing sensitive data. This approach ensures data privacy and security while benefiting from a larger and more diverse dataset and from a larger set of computing devices forming the Device–Edge–Cloud Continuum. These future directions hold significant potential for advancing occupancy prediction in buildings, enabling intelligent resource management, improved occupant experiences, and enhanced overall building performance. In the following, a detailed explanation of each highlighted point will be given.

4.1 Multi-occupancy Prediction

Occupancy prediction in buildings has garnered considerable interest due to its potential applications in energy optimization, security, and occupant comfort. However, the existing literature on occupancy prediction primarily focuses on binary occupancy scenarios, overlooking the complexities associated with multi-occupancy situations. As the number of shared spaces accommodating multiple occupants is relevant, there is a need to explore multi-occupancy prediction as a future research area within the field.

Although some research papers have been published in the field of multi-occupancy prediction in buildings [6, 47], there remains a significant gap that needs to be addressed. The exploration of multi-occupancy prediction offers numerous research opportunities. It involves developing methodologies that effectively capture and model the inter-dependencies among occupants, enabling more accurate predictions. This can be achieved through the integration of different sensors for data acquisition, advanced data analysis techniques, and the implementation of various machine learning algorithms. Deep learning approaches, such as recurrent neural networks, graph-based models, and advancements in long short-term memory neural networks, show promise in capturing the temporal dynamics and intricate interactions among occupants [35]. Furthermore, the introduction of recent technologies, including radars, smart devices, and occupant feedback, in this area can provide a comprehensive view of multi-occupancy scenarios and further enhance prediction accuracy.

4.2 Occupancy Localization

Occupancy prediction in buildings has been the subject of extensive research, and one crucial area for future investigation involves the forecasting of the localization of occupants within a building. The existing literature on occupancy prediction in buildings primarily focuses on predicting occupants' presence or absence or on multi-occupancy scenarios. However, incorporating localization can provide a deeper understanding of occupants' spatial distribution and movements.

Localization data can be leveraged to optimize energy consumption in buildings by dynamically adjusting heating, cooling, and lighting based on the occupancy distribution in different areas. Accurately forecasting occupants' locations enables building operators to implement dynamic zoning and personalized temperature control strategies. This approach focuses energy usage on providing optimal comfort exclusively in occupied areas, while unoccupied or infrequently used spaces can have reduced or deactivated HVAC operations. The targeted control of HVAC systems not only enhances occupant comfort but also results in substantial energy savings by eliminating unnecessary heating or cooling in unoccupied zones [63].

Research opportunities in occupants' localization lie in developing advanced algorithms and techniques for precise and real-time tracking within buildings. Machine learning and deep learning, along with sensor fusion techniques, can be explored to enable accurate localization by integrating data from multiple sensors. Furthermore, introducing recent technologies, i.e., UWB radar, leads to improved prediction accuracy and enhanced utilization of building resources. Accurate and reliable localization enables optimized energy management, enhanced security systems, and personalized comfort control, leading to more sustainable, secure, and user-centric built environments.

4.3 Activity Forecasting

Traditional approaches to occupancy prediction often rely on static schedules or simplistic rules, which may not accurately capture the dynamic nature of human activities within buildings. However, a promising future direction for occupancy prediction lies in activity forecasting. By leveraging historical data, contextual information, and advanced modeling techniques, activity forecasting holds the potential to revolutionize occupancy prediction in buildings [33].

Activity forecasting involves analyzing patterns, trends, and contextual cues to predict future human activities and occupancy patterns within buildings. This approach takes into account factors such as time of day, day of the week, environmental conditions, and historical occupancy data to generate accurate predictions. By incorporating activity forecasting into building management systems, numerous benefits can be realized.

The research in activity forecasting for occupancy prediction in buildings is still relatively limited [23], presenting an opportunity for researchers to explore and minimize the existing gap. To minimize the research gap in activity forecasting for occupancy prediction in buildings, the researchers can focus on data collection techniques and their analysis, develop accurate modeling techniques integrating contextual information, and validate models in real-world environments.

4.4 Federated Learning

Occupancy prediction is indeed a valuable tool for achieving energy efficiency in a building environment. However, in recent times, the collection and analysis of individual occupancy data have become challenging due to privacy concerns. FL is a promising approach to address the privacy concerns associated with occupancy prediction in building environments.

FL is a technique that allows models to be trained on decentralized datasets without the need to share data with a central server [30]. This means that data can remain private and secure while still being used to train complex machine learning models. In the context of room occupancy estimation and forecasting, where data is often collected from sensitive sources such as cameras and occupancy sensors, FL can provide a solution that balances the need for accurate data analysis with the protection of individual privacy. Moreover, FL can minimize the delay in data sharing to the cloud, as it allows the training of a model at the edge on devices that are closer to the data source. This can help to reduce network congestion and improve communication efficiency, particularly in situations where bandwidth is limited, or network latency is high. By reducing the need for data to be transmitted to a central server, FL can also minimize the risk of data breaches and protect the privacy of individuals [16].

The Device–Edge–Cloud continuum offers several advantages for occupancy prediction. First, it enables real-time analysis and decision-making at the edge, reducing the need for data transmission to the cloud for processing. This not only improves response times but also reduces the energy consumption associated with data transmission. Second, the continuum allows for a hierarchical approach to data processing, where low-level processing and filtering can be performed at the edge, while more complex analysis and model training can be conducted in the cloud. This distributed approach optimizes the use of computational resources and minimizes network bandwidth requirements.

Future research for occupancy prediction in buildings should therefore focus on exploring the potential of FL within the Device–Edge–Cloud continuum [34]. This includes investigating the development of new algorithms and models that are specifically designed to be used with FL in edge computing environments. It also involves the design of efficient communication protocols and network architectures that facilitate the seamless integration of edge devices, edge servers, and cloud resources.

5 Conclusion

This chapter has presented the current state of the art in the field of occupancy prediction in building environments by providing an overview of recent publications. The main contributions of this research were as follows.

Firstly, the exploration of IoT technologies used for monitoring presence in buildings highlighted their crucial role in assessing occupancy and facilitating efficient management of building spaces. These technologies enable the collection of data about the building's surroundings, forming the foundation for occupancy prediction.

Secondly, the examination of machine learning and deep learning algorithms showcased their significant potential in accurately forecasting the occupancy of individuals in building environments. Through the analysis of historical data and identification of patterns, these advanced computational techniques can learn and make informed predictions, thereby contributing to effective space utilization and resource allocation.

Lastly, the discussion of future directions within the field identified several areas of focus. These included enhancing the prediction accuracy of the exact number of people in buildings, integrating occupancy prediction with localization techniques, exploring activity forecasting, and investigating edge computing approaches like Federated Learning for distributed occupancy prediction.

By conducting this research, we have gained valuable insights into the advancements and opportunities in the field of occupancy prediction in building environments. The findings of this study contribute to the broader understanding of how IoT technologies and AI algorithms can be effectively utilized to optimize occupancy management, enhance energy efficiency, and improve occupant comfort and satisfaction. These insights can guide future research and practical implementations in creating smarter and more sustainable buildings.

Acknowledgments This work has been partially supported by the SoBigData.it project. SoBigData.it receives funding from European Union—NextGenerationEU—National Recovery and Resilience Plan (Piano Nazionale di Ripresa e Resilienza, PNRR)—Project: "SoBigData.it – Strengthening the Italian RI for Social Mining and Big Data Analytics"—Prot. IR0000013 – Avviso n. 3264 del 28/12/2021.

References

1. B. Abade, D. Perez Abreu, M. Curado, A non-intrusive approach for indoor occupancy detection in smart environments. Sensors **18**(11), 3953 (2018)
2. M. Abdallah, C. Clevenger, T. Vu, A. Nguyen, Sensing occupant comfort using wearable technologies, in *Construction Research Congress 2016* (2016), pp. 940–950
3. A. Abdussalam, S. Sun, M. Fu, H. Sun, I. Khan, License plate segmentation method using deep learning techniques, in *Signal and Information Processing, Networking and Computers: Proceedings of the 4th International Conference on Signal and Information Processing, Networking and Computers (ICSINC)*, 4th edn. (Springer, Berlin, 2019), pp. 58–65
4. M.S. Aliero, M.F. Pasha, A.N. Toosi, I. Ghani, The COVID-19 impact on air condition usage: a shift towards residential energy saving. Environ. Sci. Pollut. Res. **29**, 85727–85741 (2022). https://doi.org/10.1007/s11356-021-17862-z
5. G. Aloi, G. Fortino, R. Gravina, P. Pace, C. Savaglio, Simulation-driven platform for edge-based AAL systems. IEEE J. Sel. Areas Commun. **39**(2), 446–462 (2021). https://doi.org/10.1109/JSAC.2020.3021544
6. I.B. Arief-Ang, M. Hamilton, F.D. Salim, A scalable room occupancy prediction with transferable time series decomposition of CO_2 sensor data. ACM Trans. Sens. Netw. **14**(3–4), 1–28 (2018)
7. L. Atzori, A. Iera, G. Morabito, The internet of things: a survey. Comput. Netw. **54**(15), 2787–2805 (2010)
8. Z. Baird, I. Gunasekara, M. Bolic, S. Rajan, Principal component analysis-based occupancy detection with ultra wideband radar, in *2017 IEEE 60th International Midwest Symposium on Circuits and Systems (MWSCAS)* (IEEE, Piscataway, 2017), pp. 1573–1576
9. K. Bouchard, M.R. Eusufzai, R. Ramezani, A. Naeim, Generalizable spatial feature for human positioning based on Bluetooth beacons, in *2016 IEEE 7th Annual Ubiquitous Computing, Electronics & Mobile Communication Conference (UEMCON)* (IEEE, Piscataway, 2016), pp. 1–5
10. A. Bourechak, O. Zedadra, M.N. Kouahla, A. Guerrieri, H. Seridi, G. Fortino, At the confluence of artificial intelligence and edge computing in IoT-based applications: a review and new perspectives. Sensors **23**(3) (2023). https://doi.org/10.3390/s23031639. https://www.mdpi.com/1424-8220/23/3/1639
11. H. Choi, J. Lee, Y. Yi, H. Na, K. Kang, T. Kim, Deep vision-based occupancy counting: experimental performance evaluation and implementation of ventilation control. Build. Environ. **223**, 109496 (2022)
12. F. Cicirelli, A. Guerrieri, A. Vinci, G. Spezzano, *IoT Edge Solutions for Cognitive Buildings* (Springer Nature, Berlin, 2022)
13. A. Courcelle, O. Horn, Ultrasonic data representation: application to mobile robots localisation, in *Proceedings. 1998 IEEE/RSJ International Conference on Intelligent Robots and Systems. Innovations in Theory, Practice and Applications (Cat. No. 98CH36190)*, vol. 3 (IEEE, Piscataway, 1998), pp. 1559–1564
14. X. Dai, J. Liu, X. Zhang, A review of studies applying machine learning models to predict occupancy and window-opening behaviours in smart buildings. Energy Build. **223**, 110159 (2020)
15. F. Demrozi, C. Turetta, F. Chiarani, P.H. Kindt, G. Pravadelli, Estimating indoor occupancy through low-cost BLE devices. IEEE Sens. J. **21**(15), 17053–17063 (2021)
16. F. De Rango, A. Guerrieri, P. Raimondo, G. Spezzano, HED-FL: a hierarchical, energy efficient, and dynamic approach for edge federated learning. Pervasive Mob. Comput. **92**, 101804 (2023). https://doi.org/10.1016/j.pmcj.2023.101804, https://www.sciencedirect.com/science/article/pii/S1574119223000627
17. I.G. Dino, E. Kalfaoglu, O.K. Iseri, B. Erdogan, S. Kalkan, A.A. Alatan, Vision-based estimation of the number of occupants using video cameras. Adv. Eng. Inform. **53**, 101662 (2022)

18. M. Esrafilian-Najafabadi, F. Haghighat, Occupancy-based HVAC control using deep learning algorithms for estimating online preconditioning time in residential buildings. Energy Build. **252**, 111377 (2021)
19. M. Esrafilian-Najafabadi, F. Haghighat, Impact of occupancy prediction models on building HVAC control system performance: application of machine learning techniques. Energy Build. **257**, 111808 (2022)
20. N. Fatema, H. Malik, Data-driven occupancy detection hybrid model using particle swarm optimization based artificial neural network, in *Metaheuristic and Evolutionary Computation: Algorithms and Applications* (Springer, Berlin, 2020), pp. 283–297
21. M. García-Monge, B. Zalba, R. Casas, E. Cano, S. Guillén-Lambea, B. López-Mesa, I. Martínez, Is IoT monitoring key to improve building energy efficiency? Case study of a smart campus in Spain. Energy Build. **285**, 112882 (2023). https://doi.org/10.1016/j.enbuild.2023.112882
22. A. Ghosh, A. Chakraborty, D. Chakraborty, M. Saha, S. Saha, Ultrasense: a non-intrusive approach for human activity identification using heterogeneous ultrasonic sensor grid for smart home environment. J. Ambient Intell. Hum. Comput. 1–22 (2019). https://doi.org/10.1007/s12652-019-01260-y
23. K. Hasan, M. Pour Ebrahim, M.R. Yuce, Real-time people counting using IR-UWB radar, in *Proceedings of the Body Area Networks. Smart IoT and Big Data for Intelligent Health Management: 16th EAI International Conference, BODYNETS 2021, Virtual Event, October 25–26, 2021* (Springer, Berlin, 2022), pp. 63–70
24. E. Hitimana, G. Bajpai, R. Musabe, L. Sibomana, J. Kayalvizhi, Implementation of IoT framework with data analysis using deep learning methods for occupancy prediction in a building. Fut. Int. **13**(3), 67 (2021)
25. S. Hu, P. Wang, C. Hoare, J. O'Donnell, Building occupancy detection and localization using CCTV camera and deep learning. IEEE Internet Things J. **10**(1), 597–608 (2022)
26. I.I. Immoreev, P.D.V. Fedotov, Ultra wideband radar systems: advantages and disadvantages, in *2002 IEEE Conference on Ultra Wideband Systems and Technologies (IEEE Cat. No. 02EX580)* (IEEE, Piscataway, 2002), pp. 201–205
27. Z. Jiang, Z. Deng, X. Wang, B. Dong, Pandemic: occupancy driven predictive ventilation control to minimize energy consumption and infection risk. Appl. Energy **334**, 120676 (2023)
28. Y. Jin, D. Yan, X. Kang, A. Chong, S. Zhan, et al., Forecasting building occupancy: a temporal-sequential analysis and machine learning integrated approach. Energy Build. **252**, 111362 (2021)
29. D. Khalyeyev, T. Bureš, P. Hnětynka, Towards a reference component model of edge-cloud continuum, in *2023 IEEE 20th International Conference on Software Architecture Companion (ICSA-C)* (2023), pp. 91–95. https://doi.org/10.1109/ICSA-C57050.2023.00030
30. I. Khan, A. Guerrieri, G. Spezzano, A. Vinci, Occupancy prediction in buildings: an approach leveraging LSTM and federated learning, in *2022 IEEE International Conference on Dependable, Autonomic and Secure Computing, International Conference on Pervasive Intelligence and Computing, International Conference on Cloud and Big Data Computing, International Conference on Cyber Science and Technology Congress (DASC/PiCom/CBDCom/CyberSciTech)* (IEEE, Piscataway, 2022), pp. 1–7
31. A.N. Khan, A. Rizwan, R. Ahmad, D.H. Kim, An OCF-IoTivity enabled smart-home optimal indoor environment control system for energy and comfort optimization. Internet Things **22**, 100712 (2023). https://doi.org/10.1016/j.iot.2023.100712
32. S. Kim, S. Kang, K.R. Ryu, G. Song, Real-time occupancy prediction in a large exhibition hall using deep learning approach. Energy Build. **199**, 216–222 (2019)
33. K.M. Kitani, B.D. Ziebart, J.A. Bagnell, M. Hebert, Activity forecasting, in *Proceedings of the Computer Vision–ECCV 2012: 12th European Conference on Computer Vision, Florence, Italy, October 7–13, 2012*, Part IV 12 (Springer, Berlin, 2012), pp. 201–214
34. H. Kokkonen, L. Lovén, N.H. Motlagh, A. Kumar, J. Partala, T. Nguyen, V.C. Pujol, P. Kostakos, T. Leppänen, A. González-Gil, E. Sola, I. Angulo, M. Liyanage, M. Bennis, S. Tarkoma, S. Dustdar, S. Pirttikangas, J. Riekki, Autonomy and intelligence in the computing continuum: challenges, enablers, and future directions for orchestration (2023). arXiv e-prints arXiv:2205.01423

35. Y. LeCun, Y. Bengio, G. Hinton, Deep learning. Nature **521**(7553), 436–444 (2015)
36. Z. Li, B. Dong, Investigation of a short-term prediction method of occupancy presence in residential buildings, in *Proceedings of the 15th IBPSA Conference* (2017)
37. N. Li, G. Calis, B. Becerik-Gerber, Measuring and monitoring occupancy with an RFID based system for demand-driven HVAC operations. Autom. Construct. **24**, 89–99 (2012). https://doi.org/10.1016/j.autcon.2012.02.013. https://www.sciencedirect.com/science/article/pii/S0926580512000283
38. S. Matuska, R. Hudec, P. Kamencay, IoT based system for detecting the number of people and their distribution in classroom, in *2022 ELEKTRO (ELEKTRO)* (IEEE, Piscataway, 2022), pp. 1–4
39. A. Mohammadabadi, S. Rahnama, A. Afshari, Indoor occupancy detection based on environmental data using CNN-XGboost model: experimental validation in a residential building. Sustainability **14**(21), 14644 (2022)
40. H. Mohammadmoradi, S. Yin, O. Gnawali, Room occupancy estimation through WiFi, UWB, and light sensors mounted on doorways, in *Proceedings of the 2017 International Conference on Smart Digital Environment* (2017), pp. 27–34
41. I.P. Mohottige, H.H. Gharakheili, T. Moors, V. Sivaraman, Modeling classroom occupancy using data of WiFi infrastructure in a university campus. IEEE Sens. J. **22**(10), 9981–9996 (2022)
42. V. Motuzienė, J. Bielskus, V. Lapinskienė, G. Rynkun, Office building's occupancy prediction using extreme learning machine model with different optimization algorithms. Environ. Clim. Technol. **25**(1), 525–536 (2021)
43. R.C. Navarro, A.R. Ruiz, F.J.V. Molina, M.J.S. Romero, J.D. Chaparro, D.V. Alises, J.C.L. Lopez, Indoor occupancy estimation for smart utilities: a novel approach based on depth sensors. Build. Environ. **222**, 109406 (2022)
44. S. Nižetić, N. Pivac, V. Zanki, A.M. Papadopoulos, Application of smart wearable sensors in office buildings for modelling of occupants' metabolic responses. Energy Build. **226**, 110399 (2020)
45. Y. Peng, A. Rysanek, Z. Nagy, A. Schlüter, Using machine learning techniques for occupancy-prediction-based cooling control in office buildings. Appl. Energy **211**, 1343–1358 (2018)
46. A.R. Pratama, A. Lazovik, M. Aiello, Office multi-occupancy detection using BLE beacons and power meters, in *2019 IEEE 10th Annual Ubiquitous Computing, Electronics & Mobile Communication Conference (UEMCON)* (IEEE, Piscataway, 2019), pp. 0440–0448
47. B. Qolomany, A. Al-Fuqaha, D. Benhaddou, A. Gupta, Role of deep LSTM neural networks and wi-fi networks in support of occupancy prediction in smart buildings, in *2017 IEEE 19th International Conference on High Performance Computing and Communications; IEEE 15th International Conference on Smart City; IEEE 3rd International Conference on Data Science and Systems (HPCC/SmartCity/DSS)* (IEEE, Piscataway, 2017), pp. 50–57
48. S.H. Ryu, H.J. Moon, Development of an occupancy prediction model using indoor environmental data based on machine learning techniques. Build. Environ. **107**, 1–9 (2016)
49. E. Samani, P. Khaledian, A. Aligholian, E. Papalexakis, S. Cun, M.H. Nazari, H. Mohsenian-Rad, Anomaly detection in IoT-based pir occupancy sensors to improve building energy efficiency, in *2020 IEEE Power & Energy Society Innovative Smart Grid Technologies Conference (ISGT)* (IEEE, Piscataway, 2020), pp. 1–5
50. C. Savaglio, G. Fortino, A simulation-driven methodology for IoT data mining based on edge computing. ACM Trans. Internet Technol. **21**(2) (2021). https://doi.org/10.1145/3402444
51. L. Scarcello, F. Cicirelli, A. Guerrieri, C. Mastroianni, G. Spezzano, A. Vinci, Pursuing energy saving and thermal comfort with a human-driven DRL approach. IEEE Trans. Hum. Mach. Syst. 1–13 (2022). https://doi.org/10.1109/THMS.2022.3216365
52. G. Stamatescu, C. Chitu, Privacy-preserving sensing and two-stage building occupancy prediction using random forest learning. J. Sens. **2021**, 1–12 (2021)
53. D. Stjelja, J. Jokisalo, R. Kosonen, Scalable room occupancy prediction with deep transfer learning using indoor climate sensor. Energies **15**(6), 2078 (2022)

54. K. Sun, Q. Zhao, J. Zou, A review of building occupancy measurement systems. Energy Build. **216**, 109965 (2020)
55. S. Taheri, A. Razban, Learning-based CO_2 concentration prediction: application to indoor air quality control using demand-controlled ventilation. Build. Environ. **205**, 108164 (2021)
56. S. Taniguchi, G. Hangli, T. Hamada, T. Michikata, N. Koshizuka, Detecting door operations using wearable devices, in *2022 IEEE 11th Global Conference on Consumer Electronics (GCCE)* (IEEE, Piscataway, 2022), pp. 172–176
57. P.W. Tien, S. Wei, J.K. Calautit, J. Darkwa, C. Wood, A vision-based deep learning approach for the detection and prediction of occupancy heat emissions for demand-driven control solutions. Energy Build. **226**, 110386 (2020)
58. J. Vanus, R. Martinek, L. Danys, J. Nedoma, P. Bilik, Occupancy detection in smart home space using interoperable building automation technologies. Hum. Centric Comput. Inf. Sci. **12**, Article 47 (2022). https://doi.org/10.22967/HCIS.2022.12.047
59. W. Wang, J. Chen, T. Hong, Occupancy prediction through machine learning and data fusion of environmental sensing and wi-fi sensing in buildings. Autom. Construct. **94**, 233–243 (2018)
60. H. Wang, G. Wang, X. Li, An RGB-D camera-based indoor occupancy positioning system for complex and densely populated scenarios. Indoor Built Environ. **32**(6), 1198–1212 (2023). https://doi.org/10.1177/1420326X231155112
61. P.H. Winston, *Artificial Intelligence* (Addison-Wesley Longman Publishing Co., Inc., Boston, 1984)
62. Y. Yang, Y. Yuan, T. Pan, X. Zang, G. Liu, A framework for occupancy prediction based on image information fusion and machine learning. Build. Environ. **207**, 108524 (2022)
63. F. Zafari, A. Gkelias, K.K. Leung, A survey of indoor localization systems and technologies. IEEE Commun. Surv. Tutorials **21**(3), 2568–2599 (2019)

Index

C. Savaglio et al. (eds.), *Device-Edge-Cloud Continuum*, Internet of Things, https://doi.org/10.1007/978-3-031-42194-5